Study Guide for

Campbell • Reece • Simon

Essential Biology

THIRD EDITION

&

Essential Biology with Physiology

SECOND EDITION

Edward J. Zalisko, Ph.D.

Blackburn College

PEARSON

Benjamin
Cummings

San Francisco Boston New York
Cape Town Hong Kong London Madrid Mexico City Montreal Munich
Paris Singapore Sydney Tokyo Toronto

Editor-in-Chief: Beth Wilbur
Senior Acquisitions Editor: Chalon Bridges
Senior Project Manager: Ginnie Simione Jutson
Supplements Project Editor: Susan Berge
Copy Editor: Rebecca J. McDearmon/Julie Kennedy
Managing Editor: Michael Early
Supplements Production Supervisor: Jane Brundage
Marketing Managers: Christy Lawrence, Lauren Harp, Jeff Hester
Manufacturing Buyer: Stacy Wong
Production Services: Carlisle Publishing Services
Cover Image: Photograph of a "devil's flower mantis" *(Blepharopsis medica)*. Courtesy Getty Images, Inc./ Image Bank

ISBN 0-8053-0489-4

5 6 7 8—MAL—10 09 08 07
www.aw-bc.com

Contents

Preface

As you begin . . .

Have you ever felt that you studied for an exam, but your grade was still low? Have you ever wished there was a better way to study?

Ask yourself these questions:
- How do you know what to study?
- How do you know when to study?
- How do you know what you already know?
- How do you know when you are done?

If these questions leave you wondering, you might benefit from the study tips below.

Ten Tips for Studying Biology

There are many subjects and many ways to study. These are tips that have proven successful for my biology students. Try them, refine them, and tailor them to your specific needs.

1. **Read your textbook assignments before lectures over the same material.**
   Your comprehension of the material and the quality of your notes are greatly improved if you read ahead. Think about it—if you have read ahead, the lecture material will seem familiar, your notes will make more sense, and you will need to take fewer notes because you know what is included in the textbook. *Research reveals that reading ahead is one of the most important ways to improve a student's comprehension and course grades.*

2. **Study on a regular schedule.**
   Waiting until the last minute to study usually produces disappointing results. One bad grade can really pull down an average of many good grades. Students who play sports, play an instrument, or have participated in a play all know about the need for *regular practice*. We all know that daily practice is required to perform well. The same is true in the classroom.
   - Start studying when the course begins. Don't wait until a week before an exam.
   - Establish good study habits early. Don't wait until you've received poor grades.
   - Know that a little bit of studying over a long period of time is much better than the same amount of studying over a short period of time. An hour a day for 15 days is many times better than 15 hours at one time (cramming). You can spend less time than cramming and get better grades!

3. **Take frequent and short study breaks.**
   - Many times when we study, our minds drift as we listen to someone in the hallway, a sound outside, or get distracted by people around us. There are two main methods to address this problem.

- First, study in short bursts of 20–30 minutes each. Then take a 5–10 minute break to rest your mind. After the break, settle back in and focus intensely on your work. Repeat as necessary!
- Second, try to study in ways that are more interactive. The additional tips below describe a highly successful interactive way to study.

4. Identify all of the material that is addressed by the exam. Know what you need to know.

- After the first day of class, start identifying material for the first exam.
- Use your course syllabus to help you determine this material.
- If you are still unclear, talk to your instructor.

5. Try to write many questions that address all of the information for the exam.

- Write questions that can be answered with short answers, 5–20 words or so.
- Many students like to use 3×5 cards, with a question on the front and the answer on the back.
- Consider using some of the questions from the study guide.
- Begin with definitions and short lists from the material you need to study.
- Write these questions after every lecture, keeping up with the material.
- Writing the questions checks your understanding of the material. Use this opportunity to clear up confusion.

6. Quiz yourself every day or so on all of the questions you have created.

- As you try to answer the questions, create a pile of cards with questions you got correct and a pile for those questions that you missed.
- After quizzing yourself over all the questions, return to the pile of questions that you missed.
- As you quiz yourself over the questions you missed, again create a pile of those you got correct and those questions you have now missed twice.
- Look at the questions that you have missed twice. Use the Web/CD activities and your textbook to review this material again. Try to figure out ways to remember this difficult information.
- With every new lecture, add new questions to the list of questions you have already created.

7. Stop studying when you have correctly answered every question.

- Finally, a way to study that builds confidence, takes less time, and has a clear end!
- As you continue to quiz yourself, the questions will seem easier and you should feel more confident.
- Remember to keep up with your writing and quizzing. This process doesn't work well unless done regularly.

8. Use short 10–30 minute periods of your day to quiz yourself.

- Students that manage their time well will use the short periods before, between, and after classes to review.
- The result is more free time.

9. Study during the times of the day when you are most alert.

- Many students study in the evenings, when their mind and/or body is tired.
- Find times in your day when you are most alert and make these your study times.
- Use times when you tend to get sleepy for other activities (for many of us, this is early afternoon).

10. Use the night before a test to quickly review. Then get plenty of sleep.

- These techniques build confidence in your knowledge of the course material.
- Many students need only a quick review of their stack of note cards the night before the test. This might take an hour or so.
- Then get plenty of sleep. A well-rested mind always functions better.
- Review the questions that you missed on the exam and see if the information was in your note cards. Try to improve your study methods based upon the reasons for missing these questions.

Using Your Study Guide

The purpose of this study guide is to help you organize and review the information presented in the *Essential Biology* text. Each textbook chapter corresponds to a study guide chapter of the same number. The common components of each study guide chapter are described below in the sequence that they appear. Take a few minutes to review how they can help you master the textbook information.

Studying Advice

These study tips provide specific advice for best approaching each chapter.

Organizing Tables

These tables provide the framework for organizing complex information presented in the textbook. They help you organize the information for easier review.

Content Quiz

This is the largest section of each chapter. Some students use these questions to develop their note card questions. Others prefer to quiz themselves after they have mastered their note cards. The answers to the content quiz are located in the back of the study guide. The Content Quiz includes the following types of questions:

Matching—Most chapters have at least one set of matching questions which include long lists of terms or names.

Multiple Choice—These are generally short questions with one correct answer. Consider using the correct statements in these questions for note card questions.

Correctable True/False—These true/false statements have underlined words that are to be corrected if the statement is false. These corrected statements also make good material for note card questions.

Fill-in-the-Blank—These questions check your knowledge of definitions. They also make excellent note card questions.

Figure Quiz—These questions address the information presented in the textbook figures. You will need to refer to the figures in your textbook to answer them.

Analogy Questions—These questions ask you to identify relationships that are analogous to something in biology. For example, the shape of a DNA molecule is most like the shape of a spiral staircase.

Word Roots

This list presents the meanings of the word roots that form the key terms. Learning the meaning of word roots helps you guess the meaning of new terms. For example, you might know that "milli" means one thousandth and "centi" means one hundredth because you know that a millimeter is one thousandth of a meter and a centimeter is one hundredth of a meter. Therefore, you might guess that a millipede has more legs than a centipede!

Key Terms

This list consists of all the boldfaced terms and phrases in each chapter.

Crossword Puzzle

The crossword puzzle consists of the key terms and key phrases from the chapter. This is a fun way to review your knowledge of these definitions. The crossword answers are also listed at the back of the study guide.

CHAPTER 1

Introduction: Biology Today

Studying Advice

a. This introductory chapter sets the stage for the rest of the book. Here the authors focus on the concepts and less so on the terminology, so few boldfaced terms are included in this chapter. In addition, in just this first chapter, the authors have not included a chapter-ending summary or other items to help you review. The materials in this first study guide chapter are therefore of greater significance.

b. If you have not already done so, read through the introductory pages of this study guide to familiarize yourself with the typical components of each study guide chapter and review the advice for studying biology. This advice has proven important to thousands of my students in many different types of science classes. It might also be quite helpful to you.

c. In addition to using this study guide, spend time reviewing the media activities that are noted in your textbook and that appear on the CD-ROM included with your text. These activities illustrate and review significant principles addressed in your textbook.

d. Some of the multiple-choice questions included throughout this study guide rely upon a format that you might not have often encountered before. Instead of considering a long list of incorrect statements and finding the one that is correct, you are presented with a list of true statements and asked to identify the one choice that is false. (For an example of this type of format, see question 9 below.) This format is increasingly included on many types of exams, so it is helpful to practice a strategy for this style. As you consider your response to this form of multiple-choice question, consider treating each choice as a true/false question. You might even note a T or F in front of the letter identifying the choice. Your final answer will then be the one statement you determine is false.

In this study guide, this multiple-choice question format has the additional advantage of presenting a long list of true statements that can continue to be reviewed later. If you correct the single false statement, you add to this list of information for review. These single true statements also help break the information down for the construction of note card questions.

Student Media

Activities

The Levels of Life Card Game
Energy Flow and Chemical Cycling
Classification Schemes
Darwin and the Galápagos Islands
Science and Technology: DDT

Case Studies in the Process of Science

How Do Environmental Changes Affect a Population of Leafhoppers?
The Process of Science: How Does Acid Precipitation Affect Trees?

Graph It

An Introduction to Graphing

Videos

Discovery Channel Video Clip: Antibiotic Resistance
Seahorse Camouflage

You Decide

What Can We Do About Antibiotic-resistant Bacteria?

Organizing Tables

Distinguish between prokaryotic and eukaryotic cells in the table below.

TABLE 1.1		
Characteristics	Prokaryotic Cells	Eukaryotic Cells
Compare the relative size of organisms in each group.		
Indicate which groups have cells that are subdivided by internal membranes forming organelles.		
List examples of each group.		

Identify the major groups of life in the table below.

TABLE 1.2			
List the three domains of life.			
Indicate if the members of the group have prokaryotic cells or eukaryotic cells.			

Compare the groups within the domain Eukarya.

TABLE 1.3	Protista	Plantae	Animalia	Fungi
Are the organisms typically multicellular or unicellular?				
Describe their mode of nutrition.				
Indicate examples of each group.				

Content Quiz

Directions: Identify the *one* best answer for the multiple-choice questions. For true/false questions, determine if the statement is true or false. If false, change the underlined word(s) to make the statement true. Finally, add the correct word(s) to the fill-in-the-blank questions to make the statements true.

Biology and Society: Living in a Golden Age of Biology, The Scope of Life
THE UNITY OF LIFE

1. Which one of the following is *not* a property or process of life?
   A. reproduction
   B. a tendency toward disorder
   C. growth and development
   D. energy utilization
   E. evolution

2. Which one of the following is a property of life that is best illustrated by shivering and sweating?
   A. evolution
   B. order
   C. reproduction
   D. growth and development
   E. regulation

3. True or False?   Information carried by <u>proteins</u>, the units of inheritance, controls the patterns of growth and development.

4. All organisms respond to environmental _____.

LIFE AT ITS MANY LEVELS

5. Which one of the following is a correct sequence of levels of biological organization from most general to more specific?
   A. population, organism, organ system, organs, tissues
   B. organs, cells, tissues, molecules, atoms
   C. ecosystem, population, community, organism, organ system
   D. community, population, organ system, organism, organs
   E. organism, organ system, organs, cells, tissues

6. True or False?   A <u>community</u> is a group of interacting individuals of just one species.

7. Tissues are made of _____, and molecules are made of _____.

8. The dynamics of any ecosystem depend on two main processes. These are:
   A. the cycling of nutrients and the flow of energy from sunlight to producers and then to consumers and decomposers.
   B. the cycling of nutrients and the movement of water through the water cycle.
   C. the movement of water through the water cycle and photosynthesis.
   D. photosynthesis and cellular respiration.

9. Which one of the following statements comparing prokaryotic and eukaryotic cells is *false*?
   A. Eukaryotic cells are usually larger than prokaryotic cells.
   B. Eukaryotic cells are usually more complex than prokaryotic cells.
   C. Only eukaryotic cells have DNA; prokaryotes use RNA.
   D. Eukaryotic cells occur in animals and plants; bacteria have prokaryotic cells.

10. True or False?   A <u>cell</u> is the lowest level of structure that can perform all activities required for life, including the capacity to reproduce.

11. The _____, or global ecosystem, is the sum of dynamic processes in all ecosystems.

12. The branch of biology that investigates the relationships between organisms and their environments is called _____.

13. The _____ is the basic unit of structure and function.

14. The entire "book" of genetic instructions an organism inherits is called its _____.

LIFE IN ITS DIVERSE FORMS

Matching: Match the group on the left to its best description on the right.

_____ 15. Protista

_____ 16. Plantae

_____ 17. Fungi

_____ 18. Animalia

A. multicellular eukaryotes that obtain food by ingestion

B. eukaryotic organisms that are generally single-celled

C. multicellular eukaryotes that absorb nutrients by breaking down dead organisms and organic wastes

D. multicellular eukaryotes that produce their own sugars and other foods by photosynthesis

19. Which one of the following choices correctly lists the groups in their order of abundance, from the group with the greatest number of species to the group with the fewest species?
   A. plants, insects, vertebrates
   B. vertebrates, plants, insects
   C. plants, vertebrates, insects
   D. insects, plants, vertebrates
   E. insects, vertebrates, plants

20. Which of the following are two domains that include organisms with prokaryotic cells?
   A. Bacteria and Eukarya
   B. Bacteria and Archaea
   C. Archaea and Eukarya
   D. Bacteria and Fungi

21. Look at Figure 1.9 of your text. Which one of the following groups contains the smallest organisms?
   A. domain Archaea
   B. kingdom Protista
   C. kingdom Plantae
   D. kingdom Animalia
   E. kingdom Fungi

22. True or False?   All life is presently classified into <u>four</u> domains.

23. The branch of biology that names and classifies species is called _____.

24. The kingdoms of life can be assigned to even-higher levels of classification called _____.

Evolution: Biology's Unifying Theme
THE DARWINIAN VIEW OF LIFE

25. Which one of the following traits most likely occurred in the most recent common ancestor of a lizard, pigeon, turtle, and bear?
   A. fur
   B. feathers
   C. shell
   D. wings
   E. backbone

26. In his book *The Origin of Species*, Charles Darwin developed two main points. These were that:

    A. the world is very old and evolution occurs by natural selection.

    B. evolution occurs slowly and new species form by spontaneous generation.

    C. new species evolve by mutations and new species don't reproduce with old species.

    D. modern species descended from ancestral species and organisms evolve by natural selection.

27. Examine the relationships between finches in the diagram in Figure 1.13 of your text. Which one of the following pairs of finches is most closely related?

    A. warbler finches, woodpecker finch

    B. mangrove finch, woodpecker finch

    C. small tree finch, mangrove finch

    D. large cactus ground finch, small tree finch

    E. large ground finch, small tree finch

28. True or False?   The common ancestor of bears and chipmunks <u>had</u> hair and mammary glands.

29. _____ is the theme that unifies all of biology.

30. In *The Origin of Species*, Darwin proposed a mechanism for decent with modification, which he called _____.

NATURAL SELECTION

31. Which one of the following pairs of facts was the basis for Darwin's conclusion of unequal reproductive success?

    A. individual variation; offspring are the products of a blending of the parental genetics

    B. individual variation; overproduction and competition

    C. overproduction and competition; males usually fight for a mate

    D. parents produce offspring similar to themselves; overproduction and struggle for existence

    E. males usually fight for a mate; parents produce offspring similar to themselves

32. Darwin could see that, in artificial selection, humans are substituting for:

    A. individual variation.

    B. overproduction of offspring.

    C. the generation of genetic diversity.

    D. the environment.

33. In the example of the beetles discussed in the text:
    A. the most common beetle color changed after the environment changed.
    B. the environment created favorable characteristics.
    C. natural selection favored traits that worked best in different environments.
    D. the lightest beetle color pattern was most adaptive.

34. What do artificial selection and natural selection have in common?
    A. Both result from a need or desire.
    B. Both are purposeful processes.
    C. Both rely upon individual variation.
    D. Both are directional, with selection toward a goal.

35. Evolution works most like:
    A. remodeling an old home.
    B. an architect designing a new home.
    C. people in a community planning where to build a park.
    D. people voting in an election.

36. True or False?   The product of natural selection is <u>adaptation</u>.

37. Darwin used the phrase _____ to refer to unequal reproductive success.

The Process of Science
DISCOVERY SCIENCE

38. Which one of the following is an example of discovery science?
    A. comparing the effects of two different drugs on the healing of wounds
    B. testing to see if large doses of vitamin C will prevent the common cold
    C. describing the structure of a newly discovered dinosaur skull
    D. testing tropical plants for chemicals that might help fight cancer

39. True or False?   An ecologist describing all the animals and plants in a region of a tropical rain forest is using <u>hypothesis-driven</u> science.

40. The word _____ is derived from a Latin verb meaning "to know."

41. Discovery science relies upon _____ reasoning, generalizing from many observations.

HYPOTHESIS-DRIVEN SCIENCE

42. Consider the following statement. "If all vertebrates have backbones, and turtles are vertebrates, then turtles have backbones." This statement is an example of:
   A. a hypothesis.
   B. discovery science logic.
   C. rationalization.
   D. deductive reasoning.

43. True or False?   Inductive reasoning flows from the general to the specific.

44. A tentative answer to a question defines a(n) _____.

THE PROCESS OF SCIENCE: CAN COLORS PROTECT A SNAKE?

45. In the snake experiment, the statement "The coral snake–like appearance of kingsnakes repels predators" is an example of a(n):
   A. experimental question.
   B. observation.
   C. conclusion.
   D. experiment.
   E. hypothesis.

46. In the snake color experiment, the kingsnakes were using _____, an evolutionary adaptation to look like more harmful coral snakes.

47. In most cases, a control group and an experimental group differ by just one _____.

THE CULTURE OF SCIENCE; SCIENCE, TECHNOLOGY, AND SOCIETY

48. Which one of the following is *not* a characteristic of modern science?
   A. repeatability of experiments
   B. dependence upon observations
   C. requirement that ideas be testable
   D. independence and isolation of researchers

49. True or False?   Scientists usually pay <u>little</u> attention to other researchers currently working on the same problem.

50. Science and _____ are interdependent.

Evolution Connection: Theories in Science

51. Which one of the following does *not* apply to scientific theories? Scientific theories:

    A. provide a comprehensive explanation.

    B. are another name for a hypothesis.

    C. are supported by abundant evidence.

    D. are widely accepted.

52. The one theme that continues to hold all of biology together is _____.

Word Roots

hypo = below (hypothesis: a tentative explanation)

Key Terms

biology	discovery science	life	scientific method
case study	hypothesis	natural selection	theory
controlled experiment	hypothesis-driven science	science	

Crossword Puzzle

Use the Key Terms list from this chapter to fill in the crossword puzzle.

ACROSS

1. a series of steps that guide scientific investigations

4. a tentative explanation

5. a way that a population of organisms can change over generations as a result of inheritable differences in reproduction

6. a widely accepted explanatory idea supported by a large body of evidence

7. the process of seeking natural causes for natural phenomena

9. a phenomenon that includes order, regulation, growth, development, reproduction, energy utiliziation, response, and evolution

10. a type of experiment designed to compare an experimental group with a control group

11. the type of science that tests tentative explanations

DOWN

2. an in-depth examination of an actual investigation

3. a type of science that includes observations, measurements, and other descriptions

8. the scientific study of life

CHAPTER 2

Essential Chemistry for Biology

Studying Advice

a. If this is your first experience with chemistry, work slowly to master the terminology and basic relationships of atoms and their interactions. Before reading the chapter, look through the figures and read the chapter summary. This will give you an idea of the subjects that will be addressed in the text.

b. Understanding the terminology and ideas in this chapter is essential to understanding the content in the next few chapters. These next chapters continue to explore the molecules of life and how they interact to form cells. Your investment of energy in this chapter will pay dividends in the chapters to come!

c. Finally, think carefully about how you study. Refer to the study advice at the start of this study guide and establish good habits now.

Student Media

Activities

The Structure of Atoms
Electron Arrangement
Build an Atom
Ionic Bonds
Covalent Bonds
The Structure of Water
The Cohesion of Water in Trees
Acids, Bases, and pH

Case Studies in the Process of Science

How Are Space Rocks Analyzed for Signs of Life?
How Does Acid Precipitation Affect Trees?

MP3 Tutors

The Properties of Water

Videos

Discovery Channel Video Clip: Early Life

Organizing Tables

Compare the three main subatomic particles.

TABLE 2.1			
	Relative Size Compared to Other Parts of an Atom	Electrical Charge, if Any	Location in an Atom
Proton			
Neutron			
Electron			

Compare the three main types of chemical bonds.

TABLE 2.2		
	Nature of the Bond	Example
Ionic bond		
Covalent bond		
Hydrogen bond		

Describe your own examples for each of the following properties of water.

TABLE 2.3	
Property of Water	Examples from Your Life
The cohesion of water and surface tension	
How water moderates temperature	
Ice floating	
Water as a solvent	

Content Quiz

Directions: Identify the *one* best answer for the multiple-choice questions. For true/false questions, determine if the statement is true or false. If false, change the underlined word(s) to make the statement true. Finally, add the correct word(s) to the fill-in-the-blank questions to make the statements true.

Biology and Society: AIDS: A Matter of Chemistry

1. What exactly is the "molecular handshake" that is the key event in the development of AIDS?
    A. the binding of proteins on the outside of HIV to molecules on the outside of helper T cells
    B. the way that an AIDS virus is able to fold specifically inside a protein shell
    C. the very specific merger of the AIDS virus into the DNA of human white blood cells
    D. the specific binding of the AIDS virus onto the nuclear membrane of mast cells
    E. the ability of the AIDS virus to snake through tiny holes in the outside of helper T cells

2. True or False?   The AIDS virus specifically attacks human <u>B lymphocytes.</u>

3. Human suffering from AIDS is due largely to a failing _____ system.

Some Basic Chemistry
MATTER: ELEMENTS AND COMPOUNDS

4. Four elements make up 96% of the human body. Which one of the following is *not* one of those four major elements?

   A. carbon

   B. nitrogen

   C. iron

   D. oxygen

   E. hydrogen

5. True or False?   Carbon dioxide is an example of a <u>compound,</u> because it contains two or more elements in a fixed ratio.

6. Elements that your body needs in very small amounts are called _____.

ATOMS

7. Which one of the following statements about atoms is *false*?

   A. All atoms of an element have the same number of protons.

   B. Atoms whose outer shells are not full tend to interact with other atoms.

   C. An atom is the smallest unit of matter that still retains the properties of an element.

   D. A proton and an electron are almost identical in mass.

   E. The farther an electron is from the nucleus, the greater its energy.

8. The way that a satellite orbits Earth is most like the relationship between:

   A. a proton and a neutron of an atom.

   B. a cell and its molecules.

   C. an electron and the nucleus of an atom.

   D. organisms in an ecosystem.

   E. a cell in a tissue.

9. Examine Figure 2.5 of your text, which includes a helium atom. Which one of the following is a problem with the way this helium atom is represented?

   A. The protons and neutrons don't form the nucleus.

   B. Electrons are actually larger than protons and neutrons.

   C. Electrons don't orbit around the nucleus.

   D. The distance between the electrons and the nucleus is many times greater.

   E. Sometimes the electrons contribute to the structure of the nucleus.

10. True or False?   Isotopes of an element have the same number of <u>protons</u> and <u>electrons</u> but different numbers of <u>neutrons</u>.

11. The mass number is the sum of the numbers of _____ and _____.

CHEMICAL BONDING AND MOLECULES

12. When a covalent bond occurs,
    A. protons are transferred from one atom to another.
    B. the atoms in the reaction become positively charged.
    C. an atom gives up one or more electrons to another atom.
    D. ions are formed.
    E. two atoms share one or more electrons.

13. A person borrowing money from a bank is most like the relationship between:
    A. atoms that form a covalent bond.
    B. atoms that form an ionic bond.
    C. a proton and a neutron in an atom.
    D. two isotopes of an element.
    E. an organism and its cells.

14. Examine the electrons in the atoms in Figure 2.9 of your text. How many atoms of oxygen will covalently bond to one atom of carbon?
    A. 1 atom of oxygen
    B. 2 atoms of oxygen
    C. 3 atoms of oxygen
    D. 4 atoms of oxygen
    E. Oxygen will not covalently bond to carbon.

15. Water molecules are attracted to each other because of:
    A. covalent bonds.
    B. hydrogen bonds.
    C. atomic bonds.
    D. ionic bonds.

16. When we get out of a shower or bath, the surface of our skin is mostly wet because:
    A. water is being released by our skin cells.
    B. water is a polarized molecule that sticks to surfaces.
    C. water from the air condenses on our skin.
    D. our body continues to produce water.

17. True or False?   In a water molecule, the electrons spend most of their time near the <u>hydrogen atoms</u>.

18. True or False?   A molecule of carbon dioxide is a <u>compound</u> because it consists of two or more elements.

19. Water's two hydrogen atoms are joined to the oxygen atom by a(n) _____ bond.

20. Atoms that are electrically charged as a result of gaining or losing electrons are called _____.

CHEMICAL REACTIONS

21. In the chemical reaction of $2 H_2 + O_2 \rightarrow 2 H_2O$,
    A. water is a reactant.
    B. hydrogen and oxygen are the products.
    C. the number of atoms of reactants is more than the number of atoms of the products.
    D. a chemical compound is formed.

22. True or False?   Chemical reactions <u>cannot</u> destroy or create matter.

23. New chemical bonds are formed in the _____ of a reaction.

Water and Life
WATER'S LIFE-SUPPORTING PROPERTIES

24. Which one of the following statements about water is *false*?
    A. Covalent bonds give water unusually high surface tension.
    B. Water absorbs and stores a large amount of heat while warming up only a few degrees.
    C. Evaporative cooling occurs because water molecules with the greatest energy vaporize first.
    D. Ice floats because water molecules move farther apart when they form a solid.
    E. Water is a common solvent inside your body.

25. Which one of the following is the best example of surface tension?
    A. floating ice
    B. a water strider "standing" on the surface of liquid water
    C. sweat cooling the surface of your skin
    D. sugar dissolving quickly in a cup of hot tea
    E. a squirrel standing on an ice-covered surface of a pond

26. It is a hot summer day and Jill just finished jogging. As she grabs a drink, ice cubes floating at the top of the glass hit her nose and lemonade spills out. She uses her towel to wipe up drops of lemonade clinging to her cheeks. Which one of the following properties of water was *not* described in this situation?

    A. evaporative cooling
    B. the cohesive properties of water
    C. the ability of ice to float
    D. All of the above were noted in the description.

27. Which one of the following is represented in Figure 2.13 of your text?

    A. surface tension
    B. ability of water to moderate temperature
    C. ability of ice to float
    D. evaporative cooling
    E. versatility of water as a solvent

28. Your instructor calculates the average score on the last exam. Then your instructor eliminates the top three scores and recalculates the average, which is now quite a bit lower. This recalculation is most like the way that:

    A. sweat cools your skin through evaporative cooling.
    B. ice is able to expand, permitting it to float.
    C. molecules of water release energy to form surface tension.
    D. solutes are broken down when dissolved in a solvent.

29. True or False?   Because of <u>hydrogen bonding</u>, water has a better ability to resist temperature change than most other substances on Earth.

30. A(n) _____ is dissolved in a solvent to produce a(n) _____.

31. When _____ is the solvent, the result is an aqueous solution.

ACIDS, BASES, AND pH

32. The pH of a solution:

    A. inside most living cells is close to 5.
    B. can be quickly changed by the addition of a buffer.
    C. can range from 0 to 14.
    D. is acidic above the level of 7.
    E. is determined by the relative amount of OH^- ions.

33. A solution with a pH of 7 contains:
    A. more H^+ than OH^- ions.
    B. more OH^- than H^+ ions.
    C. only H^+ ions.
    D. only OH^- ions.
    E. equal amounts of H^+ and OH^- ions.

34. Which one of the following people functions most like a buffer?
    A. a comedian who keeps the audience laughing
    B. a funeral director who helps people in times of grief
    C. a counselor who tries to understand the hidden meaning in your dreams
    D. a coach pumping up a losing team and challenging a winning team
    E. an actor playing a serious role in a drama

35. True or False?   A solution with a pH of 6 has <u>100 times</u> more H^+ than an equal amount of a solution with a pH of 5.

36. Most biological fluids contain _____, substances that resist changes in pH.

37. A chemical compound that releases hydrogen ions in a solution is called a(n) _____. A chemical compound that accepts hydrogen ions and removes them from a solution is called a(n) _____.

Evolution Connection: Earth Before Life

38. Life first evolved about 3.5–4.0:
    A. billion years ago, in an atmosphere similar to what we see today.
    B. million years ago, in an atmosphere similar to what we see today.
    C. billion years ago, in an atmosphere different from what we see today.
    D. million years ago, in an atmosphere different from what we see today.

39. True or False?   The densest layers of Earth are at the <u>surface</u>.

40. The second atmosphere of Earth was formed by gases released from _____.

Word Roots

aqua = water (aqueous: a type of solution in which water is the solvent)

co = together; **valent** = strength (covalent bond: an attraction between atoms that share one or more pairs of outer-shell electrons)

iso = equal (isotope: an element having the same number of protons and electrons but a different number of neutrons)

neutr = neither (neutron: a subatomic particle with a neutral electrical charge)

Key Terms

acid
aqueous solution
atom
atomic number
base
buffer
chemical bond
chemical reaction
cohesion
compound

covalent bond
electron
element
evaporative
  cooling
heat
hydrogen bond
ion
ionic bond

isotope
mass
mass number
matter
molecule
neutron
nucleus
pH scale
polar molecule

product
proton
radioactive isotope
reactant
solute
solution
solvent
temperature
trace element

Crossword Puzzle

Use the Key Terms list from this chapter to fill in the crossword puzzle.

ACROSS

3. a fluid mixture of two or more substances

6. a starting material in a chemical reaction

8. a type of isotope whose nucleus decays spontaneously

9. a subatomic particle that is electrically neutral

10. a substance containing two or more elements in a fixed ratio

12. an attraction between atoms that share one or more pairs of electrons

13. a variant form of an atom

14. an atom or molecule that has gained or lost one or more electrons

17. a substance that increases the hydrogen ion concentration in a solution

18. the amount of energy associated with the movement of the atoms and molecules in a body of matter

20. anything that occupies space and has mass

22. a molecule that has opposite charges on opposite ends

24. the attraction between molecules of the same kind

26. a substance that cannot be broken down into other substances

28. a measure of the amount of material in an object

29. an attraction between two atoms

30. the smallest unit of matter that retains the properties of an element

31. a chemical substance that resists changes in pH

32. the sum of the number of protons and neutrons in an atom's nucleus

34. a weak chemical bond between a partially positive hydrogen and a partially negative atom

35. surface cooling that results when a substance evaporates

DOWN

1. an element that is essential for the survival of an organism but only in minute quantities

2. the dissolving agent in a solution

4. an atom's central core

5. a subatomic particle with a single positive electrical charge

7. an attraction between two ions with opposite electrical charges

10. a process leading to chemical changes in matter

11. a measure of the relative acidity of a solution, ranging in value from 0 to 14

15. a measure of the intensity of heat

16. a substance that is dissolved in a solution

19. an ending material in a chemical reaction

21. a mixture of two or more substances, one of which is water

23. a group of two or more atoms held together by covalent bonds

25. the number of protons in each atom of a particular element

27. a subatomic particle with a negative charge

33. a substance that decreases the hydrogen concentration in a solution

The Molecules of Life

Studying Advice

a. Chapter 3 introduces the four main types of biological molecules. The organizing table below will help you compare these categories and organize the details of each group.

b. Find a nutrition label on some packaged food around you. Notice that three of the four categories of biological molecules discussed in this chapter are sources of nutrition. These three types—carbohydrates, proteins, and lipids—are all sources of calories. The chemistry in this chapter is something you already know a little about.

Student Media

Activities

Diversity of Carbon-Based Molecules
Functional Groups
Making and Breaking Polymers
Models of Glucose
Carbohydrates
Lipids
Protein Functions
Protein Structure
Nucleic Acid Functions
Nucleic Acid Structure

Case Studies in the Process of Science

What Factors Determine the Effectiveness of Drugs?

MP3 Tutors

Protein Structure and Function

You Decide

Low-fat or Low-carb Diets—Which is Healthier?

Organizing Tables

Compare the four classes of large biological molecules by completing the table below. Some of the cells are already filled in to help you complete the table.

TABLE 3.1			
Group	Monomer	Subgroups	Examples
Carbohydrates		Monosaccharides	
		Disaccharides	
		Polysaccharides	
Lipids	*No Monomer*	Unsaturated fats	
		Saturated fats	
		Steroids	
Proteins		*No Subgroup*	
Nucleic acids		DNA	
		RNA	

Content Quiz

Directions: Identify the *one* best answer for the multiple-choice questions. For true/false questions, determine if the statement is true or false. If false, change the underlined word(s) to make the statement true. Finally, add the correct word(s) to the fill-in-the-blank questions to make the statements true.

Biology and Society: Does Thanksgiving Dinner Make You Sleepy?

1. A typical Thanksgiving meal is rich in:
   A. carbohydrates.
   B. proteins.
   C. fats.
   D. all of the above.
   E. none of the above.

2. True or False?   Fish contains more tryptophan by weight than is found in turkey.

3. Once digested, tryptophan can be converted to _____, a chemical that can act on the brain to promote sleep.

Organic Molecules
CARBON MOLECULES

4. Which one of the following statements about organic molecules is *false*?
    A. It is possible to construct an endless diversity of carbon skeletons.
    B. The simplest hydrocarbon is methane.
    C. Carbon can use only one bond to attach to another carbon atom.
    D. Carbon completes its outer shell by sharing electrons with up to four other atoms.
    E. Carbon frequently bonds to the elements hydrogen, oxygen, and nitrogen.

5. Examine the bonds between the carbons in Figure 3.2 of your text. Find a pair of carbons with a double line between them. The double line indicates that these two carbons:
    A. share two electrons.
    B. share two hydrogens.
    C. are joined by an ionic bond.
    D. are joined at the nucleus.
    E. share electrons with hydrogens that we cannot see.

6. True or False?   The groups of atoms that usually participate in chemical reactions are called <u>functional groups</u>.

7. In methane, carbon is joined to four hydrogen atoms by _____ bonds.

GIANT MOLECULES FROM SMALLER BUILDING BLOCKS

8. Which one of the following processes is the reverse of a dehydration reaction?
    A. hydrolysis
    B. hydration
    C. denaturation
    D. diffusion
    E. osmosis

9. True or False?   In the process of a dehydration reaction, a molecule of <u>carbon dioxide</u> is formed.

10. In the same way that a train is made by linking together railroad cars, cells make _____ by linking together _____.

Biological Molecules
CARBOHYDRATES

11. Automobiles use gasoline the way that cells use:
    A. cellulose.
    B. amino acids.
    C. glucose.
    D. proteins.
    E. lipids.

12. Sucrose is:
    A. the main carbohydrate in plant sap.
    B. the main sweetener in soft drinks.
    C. a monosaccharide.
    D. sweeter tasting than fructose.
    E. abundant in corn syrup.

13. Plants use starch the way that animals use:
    A. glucose.
    B. glycogen.
    C. fructose.
    D. cellulose.
    E. sucrose.

14. Compare the structures of glucose and fructose in Figure 3.8 of your text. How are these molecules different?
    A. Glucose has more double bonds.
    B. Fructose has more carbon atoms.
    C. Fructose has more hydrogen atoms.
    D. Glucose has more oxygen atoms.
    E. Only the shapes are different.

15. Examine the structure of glucose in Figure 3.9 of your text. How are the linear and ring structures different?
    A. The linear structure has more carbon atoms.
    B. The linear structure has more oxygen atoms.
    C. The ring structure has more hydrogen atoms.
    D. The ring structure has more double bonds.
    E. Only the shapes are different.

16. True or False?   Starch is the most abundant organic compound on Earth.

17. True or False?   Cellulose cannot be hydrolyzed by most animals.

18. Glucose and fructose are examples of _____, molecules that match in their molecular formulas but differ in their shapes.

19. Cells construct a disaccharide by joining two _____ in a(n) _____ reaction.

20. Starch is a polysaccharide made by joining together many _____ monomers.

LIPIDS

21. Unsaturated fats:
    A. are usually solid at room temperature.
    B. contribute to cardiovascular disease more than saturated fats.
    C. have the maximum number of hydrogen atoms attached.
    D. lack double bonds in their hydrocarbon portions.
    E. none of the above.

22. If we add hydrogen to an unsaturated fat, we would expect that the unsaturated fat would:
    A. have more double bonds.
    B. be a much longer molecule.
    C. be stiffer at room temperature.
    D. likely be healthier.
    E. none of the above.

23. Which one of the following statements about cholesterol is *false*? Cholesterol is:
    A. made of a carbon skeleton consisting of four fused rings.
    B. one of the most well-known steroids.
    C. used by our bodies to make sex hormones.
    D. similar to fats in structure and function.
    E. present in cell membranes.

24. Anabolic steroids:
    A. structurally resemble testosterone.
    B. can cause serious mental problems.
    C. are synthetic variants of testosterone.
    D. can damage the liver and lead to infertility.
    E. all of the above.

25. True or False?   A pound of fat contains <u>more than twice</u> as much energy as a pound of starch.

26. The accumulation of lipid-containing deposits within the walls of blood vessels leads to a type of cardiovascular disease called _____.

27. Lipids are _hydrophobic_, which means that they cannot mix with water.

PROTEINS

28. Which one of the following statements about proteins is *false*?
    A. A protein's three-dimensional shape enables it to carry out its normal functions.
    B. All proteins are made from a common set of just 20 amino acids.
    C. Amino acids in a protein are linked together by peptide bonds.
    D. Enzymes are specialized types of proteins.
    E. Proteins are made from nucleic acid monomers.

29. Which one of the following is most like the process of making proteins?
    A. weaving hair into a braid
    B. sewing a quilt from hundreds of patches of old clothing
    C. making a train by connecting together 20 different types of railroad cars
    D. gathering peas onto a spoon as you prepare to eat them
    E. preparing a stew by cutting up four different vegetables and mixing them with meat

30. The way a protein changes shape when it is heated is most like:
    A. tearing sections off a roll of toilet paper.
    B. cutting a piece of string into two.
    C. separating a few cars from a long train.
    D. untangling a garden hose.

31. Examine Figure 3.24 of your text to best understand the levels of protein structure. Which one of the following is most like the secondary level of protein structure?
    A. the sequence of rail cars on a train
    B. a coiled spring
    C. the sequence of letters that form a word
    D. the links of chain forming a necklace

32. True or False? <u>All amino acids</u> have a carboxyl group, an amino group, and a hydrogen atom attached to a central carbon atom.

33. The specific amino acid sequence of a protein is the protein's _primary_ structure.

34. If a protein is exposed to a change in temperature or pH, it can lose its normal shape in a process called _____.

35. Which one of the following statements about nucleic acids is *false*?
    A. The base is the part that varies between nucleotides of DNA.
    B. There are two types of nucleic acids, DNA and RNA.
    C. Nucleic acids are linked into long strands to form nucleotides.
    D. The sequence of amino acids in proteins is determined by the sequence of nucleotides in DNA.
    E. Molecules of DNA form a double helix.

36. The overall shape of a DNA molecule is most similar to the shape of:
    A. railroad tracks.
    B. a spiral staircase.
    C. a screwdriver.
    D. a bicycle chain.
    E. the number "8."

37. DNA and RNA are alike in that both:
    A. are usually double-stranded molecules.
    B. use the bases adenine, guanine, cytosine, and thymine.
    C. use the sugar deoxyribose.
    D. are polymers of nucleotides.
    E. none of the above.

38. True or False?   The genetic material humans and other organisms pass from one generation to the next consists of <u>RNA</u>.

39. Nucleic acids are made of monomers called _____.

Evolution Connection: DNA and Proteins as Evolutionary Tape Measures

40. True or False?   Scientists expect that the DNA of closely related species will be <u>more</u> similar than the DNA of distantly related species.

41. The DNA sequences determine the _____ sequences in proteins.

Word Roots

di = two (disaccharide: two monosaccharides joined together)

glyco = sweet (glycogen: a polysaccharide sugar used to store energy in animals)

hydro = water (hydrolysis: breaking chemical bonds by adding water)

iso = equal (isomers: molecules with similar molecular formulas but different structures)

lyse = break (hydrolysis: breaking chemical bonds by adding water)

macro = big (macromolecule: a giant molecule in living organisms)

meros = part (polymer: a chain made from smaller organic molecules)

mono = single (monosaccharide: simplest type of sugar)

philic = loving (hydrophilic: water-loving property of a molecule)

phobos = fearing (hydrophobic: water-hating property of a molecule)

poly = many (polysaccharide: many monosaccharides joined together)

sacchar = sugar (monosaccharide: simplest type of sugar)

sclero = hard (atherosclerosis: hardening of the arteries)

tri = three (triglyceride: a glycerol molecule joined with three fatty acid molecules)

Key Terms

amino acids
anabolic steroids
atherosclerosis
carbohydrates
cellulose
dehydration
  reaction
denaturation
disaccharides
DNA
double helix
fats

functional groups
glycogen
hydrocarbons
hydrogenation
hydrolysis
hydrophilic
hydrophobic
isomers
lipids
low-carb diets
macromolecules
monomers

monosaccharides
nitrogenous base
nucleic acids
nucleotides
organic chemistry
organic
  compounds
peptide bond
polymers
polypeptide
polysaccharides
primary structure

protein
protein shape
RNA
saturated
starch
steroids
sugar-phosphate
  backbone
trans fats
triglyceride
unsaturated

Crossword Puzzle

Use the Key Terms list from this chapter to fill in the crossword puzzle.

ACROSS

1. a type of diet that avoids starches and sugars
3. type of bond between adjacent amino acids
4. water-hating property of a molecule
5. first level of protein structure
6. unhealthy unsaturated fats produced by hydrogenation
8. molecules with similar formulas but different structures
9. the addition of hydrogens to unsaturated fats
11. the type of backbone forming polynucleotides
12. a polysaccharide used to store energy in plants
13. the simplest organic molecule
16. monomer of a nucleic acid
18. a chain of amino acids used to make a protein
20. type of group of atoms usually involved in chemical reactions
23. the part of the molecule that varies between different types of nucleotides
25. gigantic molecule
26. a change in the shape of a protein
27. simplest type of sugar
29. cardiovascular disease in which lipid deposits accumulate in walls of arteries
30. the type of chemistry that studies organic molecules
31. a type of nucleic acid that uses the base uracil
32. monomer of a protein
34. the most abundant organic compound on Earth
35. macromolecule made of one or more polypeptides

36. type of organic molecule that includes sugars and starch
37. a large lipid molecule made from glycerol and three fatty acids
38. the genetic material that organisms inherit from their parents
39. shape of DNA

DOWN

1. a type of organic molecule that includes fats, steroids, and phospholipids
2. a type of lipid with a carbon skeleton in the form of four fused rings
4. water-loving property of a molecule
7. type of fatty acid with the maximum number of hydrogens
10. process that links monomers together
14. many monosaccharides joined together
15. a polysaccharide used to store energy in animals
17. breaking chemical bonds by adding water
19. a chain made from smaller organic molecules
21. building block of a polymer
22. type of organic molecule including DNA and RNA
24. synthetic variant of testosterone
28. two monosaccharides joined together
33. type of fatty acid with less than the maximum hydrogens

CHAPTER 4

A Tour of the Cell

Studying Advice

a. Before you begin reading this chapter, examine Figure 4.6 of your text. How many of these cell parts do you recognize from prior courses? Also refer to Figure 4.3 of your text to become familiar with the relative sizes of cells and their parts.

b. The organizing tables below will be especially useful in this chapter. Many basic parts of cells are introduced. These tables will help you keep the information organized for review.

Student Media

Activities

Metric System Review
Prokaryotic Cell Structure and Function
Comparing Cells
Build an Animal Cell and a Plant Cell
Membrane Structure
Overview of Protein Synthesis
The Endomembrane System
Build a Chloroplast and a Mitochondrion
Cilia and Flagella
Review: Animal Cell Structure and Function
Review: Plant Cell Structure and Function

Case Studies in the Process of Science

What Is the Size and Scale of Our World?

eTutors

A Tour of the Cell

Videos

Discovery Channel Video Clip: Cells

Cytoplasmic Streaming

Paramecium Cilia

Paramecium Vacuole

Organizing Tables

Compare the features of prokaryotic and eukaryotic cells. (See your text and Figure 4.3 of your text.)

TABLE 4.1

	Prokaryotic Cells	Eukaryotic Cells
Which group has a nucleus bordered by a membrane?		
Which group has organelles?		
How do the sizes compare?		
What are examples of each group?		

Compare the structure, location, and function of the following components of the endomembrane system.

TABLE 4.2

	Structure	Location in the Cytoplasm	Function(s)
Smooth endoplasmic reticulum			
Rough endoplasmic reticulum			
Golgi apparatus			
Lysosome			
Central vacuole			

Compare the features of cilia and flagella.

TABLE 4.3		
	Cilia	Flagella
Which are usually longer?		
Which are usually more numerous?		
How does their basic architecture compare?		
Where are they found?		

Compare the structures and functions of mitochondria and chloroplasts.

TABLE 4.4		
	Mitochondria	Chloroplasts
These organelles are found in the cells of organisms in what kingdoms?		
Which organelle is involved in cellular respiration and which is involved in photosynthesis?		
Compare the number of membranes and compartments found in each organelle.		
Consider drawing a general sketch to show these parts.		
What part of each organelle is the most active region?		

Content Quiz

Directions: Identify the *one* best answer for the multiple-choice questions. For true/false questions, determine if the statement is true or false. If false, change the underlined word(s) to make the statement true. Finally, add the correct word(s) to the fill-in-the-blank questions to make the statements true.

Biology and Society: Cells That Cure

1. Cell therapy to repair damage caused by a heart attack:
   A. genetically modifies muscle cells surrounding the damage to promote the replacement of the damaged cells.
   B. uses a series of drugs to encourage the damaged cells to grow and reproduce to repair the damaged region.
   C. transplants into the damaged heart region mouse stem cells that have been changed into heart muscle.
   D. injects immature muscle cells from another area of the body into the damaged areas of the heart.
   E. none of the above.

2. True or False?   Even though many patients respond positively to cell therapy to treat damage caused by a heart attack, scientists <u>do not</u> yet understand how cell therapy works.

3. True or False?   Heart muscle cells <u>regenerate</u> over time.

4. During a heart attack, heart muscle cells die because they are starved for
   _____.

The Microscopic World of Cells

5. Which one of the following is most closely associated with the term "resolution"?
   A. clarity
   B. larger size
   C. greater color
   D. lighter image

6. Examine Figure 4.6 of your text. Which one of the following organelles is physically connected to the nucleus?
   A. mitochondrion
   B. Golgi apparatus
   C. rough and smooth endoplasmic reticulum
   D. lysosome
   E. plasma membrane

7. Which one of the following statements about prokaryotic and eukaryotic cells is *false*?
   A. Eukaryotic cells have membrane-enclosed organelles.
   B. Prokaryotic cells evolved before eukaryotic cells evolved.
   C. Eukaryotic cells are generally smaller than prokaryotic cells.
   D. Eukaryotic cells divide the labor of life among many internal compartments.
   E. Most bacteria are surrounded by a rigid cell wall.

8. Which one of the following has prokaryotic cells?
   A. a mushroom
   B. a bacterium
   C. an oak tree
   D. a crawfish
   E. a virus

9. True or False?   Bacteria and archaea both consist of <u>prokaryotic</u> cells.

10. True or False?   The genetic material of eukaryotic cells is housed in the <u>endoplasmic reticulum</u>.

11. Small parts of eukaryotic cells with specific functions are called _____.

12. The magnification of a transmission electron microscope is about _____ times greater than that of a typical light microscope.

13. Cell surfaces are best revealed by a(n) _____ electron microscope.

14. Internal details of cells are best revealed by a(n) _____ electron microscope.

15. A nucleus bordered by a membranous envelope is found in _____ cells, but not in _____ cells.

16. The generalization that all living things are composed of cells is called _____ theory.

17. Some prokaryotic cells have _____, which help attach them to surfaces.

18. In a eukaryotic cell, the region between the nucleus and the plasma membrane is the _____.

19. In a eukaryotic cell, organelles are suspended in a fluid called the _____.

Membrane Structure

20. The membranes of cells are primarily composed of:
    A. lipids and nucleic acids.
    B. proteins and nucleic acids.
    C. lipids and carbohydrates.
    D. proteins and carbohydrates.
    E. lipids and proteins.

21. Which one of the following statements about cellular membranes is *false?*
    A. Membrane molecules can move freely past one another.
    B. Most membranes have carbohydrates embedded in the phospholipid bilayer.
    C. Diverse proteins float like icebergs in the phospholipid sea.
    D. The structure of the plasma membrane is similar to the structure of the other internal membranes of eukaryotic cells.
    E. Phospholipids in the membrane form a bilayer.

22. Which one of the following statements about plant cell walls is *false?* Plant cell walls:
    A. help the cells maintain their shape.
    B. protect the cells from physical damage.
    C. restrain plant cells from absorbing too much water.
    D. are the primary site of photosynthesis.

23. True or False?   Animal cells are often bound to the extracellular matrix by surface <u>lipids</u> in the plasma membrane.

24. The phospholipids and most of the proteins in a membrane are free to drift about in what is called the fluid _____.

25. The surfaces of most animal cells contain cell _____, structures that connect them to another cell.

The Nucleus and Ribosomes: Genetic Control of the Cell

26. Which one of the following statements about the nucleus and ribosomes of eukaryotic cells is *false?*
    A. Chromosomes are composed of long strands of DNA attached to certain proteins.
    B. Pores in the nuclear envelope allow messenger RNA to move from the nucleus to the cytoplasm.
    C. Ribosomes are constructed in the cytoplasm from parts produced in the nucleolus.
    D. DNA moves from the nucleus to the cytoplasm to direct the production of proteins.
    E. Ribosomes may work either suspended in the cytosol or attached to the endoplasmic reticulum.

27. If we think of the cell as a factory, then the nucleus is its executive boardroom and the top managers are the:

   A. genes.

   B. ribosomes.

   C. chromosomes.

   D. mitochondria.

   E. endoplasmic reticulum.

28. True or False?   The directions to make a protein move from <u>DNA</u> to <u>messenger RNA</u> and then to <u>ribosomes</u>.

29. Proteins are made on ribosomes in the cytoplasm using the directions from a(n) _____ molecule.

30. Differences in the structures of _____ allow humans to use antibiotics that do not harm human cells.

The Endomembrane System: Manufacturing and Distributing Cellular Products

31. Which one of the following statements about the components of the endomembrane system is *false*?

   A. Products from the endoplasmic reticulum are modified in the Golgi apparatus.

   B. Lysosomes have several types of digestive functions.

   C. Lipids are typically produced by the rough endoplasmic reticulum.

   D. Central vacuoles contribute to plant growth and may contain pigments and poisons.

   E. The smooth endoplasmic reticulum helps detoxify poisons.

32. Examine Figures 4.11–4.15 of your text. Which one of the following sequences best represents the steps an enzyme would take from initial production to joining with a food vacuole?

   A. rough endoplasmic reticulum, Golgi apparatus, lysosome

   B. Golgi apparatus, rough endoplasmic reticulum, lysosome

   C. lysosome, smooth endoplasmic reticulum, Golgi apparatus

   D. smooth endoplasmic reticulum, Golgi apparatus, lysosome

33. The smooth endoplasmic reticulum:

   A. synthesizes lipids and helps to detoxify drugs.

   B. synthesizes carbohydrates and proteins.

   C. is the primary site of ribosome production.

   D. regulates the import of drugs and hormones.

   E. helps to destroy lysosomes.

34. True or False?   The final destination of some proteins in the cell is determined by chemical tags applied by the <u>rough endoplasmic reticulum</u>.

35. Found in animal cells but absent from most plant cells, _____ are membrane-bounded sacs of digestive enzymes.

36. Membrane proteins and secretory proteins are the typical products of the _____ endoplasmic reticulum.

Chloroplasts and Mitochondria: Energy Conversion

37. In a plant cell, where is light energy trapped and converted to chemical energy?
    A. the inner mitochondrial membrane
    B. the outer mitochondrial membrane
    C. the outer chloroplast membrane
    D. the stroma of a chloroplast
    E. the grana in the chloroplast

38. Compare Figures 4.16 and 4.17 of your text. Which of the following pairs of structures are the largest membranes in chloroplasts and mitochondria?
    A. outer chloroplast membrane and outer mitochondrial membrane
    B. inner chloroplast membrane and outer mitochondrial membrane
    C. outer chloroplast membrane and inner mitochondrial membrane
    D. the grana of chloroplasts and the outer mitochondrial membrane
    E. the grana of chloroplasts and the inner mitochondrial membrane

39. True or False?   Plant cells have <u>chloroplasts and mitochondria</u>.

40. Cells use molecules of _____ as the direct energy source for most of their work.

41. The extensive folds of the inner mitochondrial membrane are called _____.

The Cytoskeleton: Cell Shape and Movement

42. Which one, if any, of the following is *not* a function of the cytoskeleton?
    A. gives mechanical support to the cell
    B. anchors organelles
    C. helps a cell maintain its shape
    D. helps a cell move
    E. All of the above are functions of the cytoskeleton.

43. Which one of the following statements about the cilia and flagella is *false*?
    A. Cilia are generally shorter than flagella.
    B. Cilia are more numerous than flagella.
    C. Cilia and flagella have the same basic architecture.
    D. Flagella line the inside of your windpipe.
    E. Human sperm rely on flagella for movement.

44. True or False? A specialized arrangement of <u>microfilaments</u> helps move cilia and flagella.

45. The movement of dividing chromosomes is guided by _____.

Evolution Connection: The Origin of Membranes

46. One of the first steps in the evolution of the first cells was likely the origin of:
    A. mitochondria.
    B. membranes.
    C. chloroplasts.
    D. photosynthesis.
    E. DNA.

47. The first cell membranes allowed cells to:
    A. regulate their chemical exchanges with the environment.
    B. form an impermeable barrier between the inside and outside of the cell.
    C. use photosynthesis.
    D. reproduce.
    E. make ATP.

48. True or False? Phospholipids require <u>no</u> genetic information to assemble into membranes.

49. The main lipids in the earliest cell membranes were probably _____.

Word Roots

chloro = green (chloroplast: the green organelle of photosynthesis)

chromo = color (chromosome: a thread-like, darkly staining structure packaging DNA in the nucleus)

cili = small hair (cilium: a short, hair-like cellular appendage with a microtubule core)

cyto = cell (cytoplasm: cell region between the nucleus and the plasma membrane)

endo = inner (endomembrane system: an internal system of membranous organelles)

eu = true (eukaryotic: cell type with a membrane-enclosed nucleus and other organelles)

extra = outside (extracellular: the substance around animal cells)

flagell = whip (flagellum: a long, whip-like cellular appendage that moves cells)

micro = small (microtubules: microscopic tubular filaments contributing to the cytoskeleton)

plasm = molded (plasma membrane: the thin layer that sets a cell apart from its surroundings)

pro = before (prokaryotic: the first cells, lacking a membrane-enclosed nucleus and other organelles)

reticul = network (endoplasmic reticulum: membranous network where proteins are produced)

trans = across (transport vesicles: membranous spheres that move materials across a cell)

vacu = empty (vacuole: sac that buds from the ER, Golgi apparatus, or plasma membrane)

Key Terms

cell junctions
cell theory
central vacuole
chloroplast
chromatin
chromosome
cilia
cristae
cytoplasm
cytoskeleton
cytosol
electron microscope (EM)
endomembrane system

endoplasmic reticulum (ER)
eukaryotic cell
extracellular matrix
flagella
fluid mosaic
food vacuoles
gene
Golgi apparatus
grana
light microscope (LM)
lysosomal storage disease

lysosome
magnification
matrix
microtubule
mitochondria
nuclear envelope
nucleolus
nucleus
organelles
phospholipid
phospholipid bilayer
plasma membrane
prokaryotic cell
resolving power

ribosome
rough ER
scanning electron microscope (SEM)
smooth ER
stroma
transmission electron microscope (TEM)
transport vesicles
vacuole

Crossword Puzzle

Use the Key Terms list from this chapter to fill in the crossword puzzle.

ACROSS

1. a type of cell found only in the bacteria and archaea

4. a double membrane, perforated with pores, that encloses the nucleus

6. the thickest of the three main kinds of fibers making up a eukaryotic cytoskeleton

7. a membranous network of tubules not associated with ribosomes in a eukaryotic cell

9. a digestive organelle in eukaryotic cells

10. the organelles in eukaryotic cells where cellular respiration occurs

12. an increase in the apparent size of an object

13. a network of interconnected membranous sacs studded with ribosomes

16. initials for a type of microscope that uses an electron beam to study a specimen's surface

18. the thick fluid within the chloroplast

19. a type of cell that has a membrane-enclosed nucleus and organelles

21. interconnected stacks where sunlight is captured during photosynthesis

23. a discrete unit of hereditary information

24. a type of system of membranous organelles subdividing eukaryotic cells

28. a description of the flowing nature of membrane structure

29. initials for an instrument that focuses an electron beam to see fine details

30. a membrane-enclosed sac in a eukaryotic cell with diverse functions

32. a sticky coat secreted by most animal cells

33. a measure of the clarity of an image

34. a type of theory that all living things are composed of cells and all cells come from other cells

36. the combination of DNA and proteins that constitutes eukaryotic chromosomes

37. a membrane-enclosed sac occupying most of the interior of a mature plant cell

39. a photosynthetic organelle found in plants and some protists

40. the place in a eukaryotic cell nucleus where ribosomes are made

41. a thread-like, gene-carrying structure found in the nucleus of a eukaryotic cell

42. structures with specialized functions within a cell

43. the semifluid medium of a cell's cytoplasm

44. a membrane fat with a hydrophilic "head" and two hydrophobic "tails"

45. infoldings of the inner mitochondrial membrane

DOWN

2. membranous spheres that bud from the endoplasmic reticulum

3. initials for a type of microscope that uses a penetrating electron beam to study internal cellular details

5. a type of hereditary disease associated with abnormal lysosomes

8. a sticky coat outside of cells

11. an organelle that functions as the site of protein synthesis in the cytoplasm

14. the simplest type of digestive cavity, found in protists

15. similar to cilia in structure, these cellular appendages propel protists and sperm

17. the thin layer of lipids and proteins that sets a cell off from its surroundings

20. a network of fine fibers that provides structural support for a eukaryotic cell

22. a double layer that makes up the basic fabric of biological membranes

25. initials for a membranous organelle connected to the outer nuclear membrane

26. everything inside a cell between the plasma membrane and the nucleus

27. an organelle that modifies, stores, and ships products of the endoplasmic reticulum

31. a type of connection between cells

35. initials for an optical instrument with lenses that bend visible light to magnify images

38. an atom's central core, containing protons and neutrons

41. cellular appendages similar to flagella that move some protists through water

CHAPTER 5

The Working Cell

Studying Advice

a. This chapter addresses many abstract ideas that may initially be difficult to grasp. Do not try to understand all of this chapter in a single night. Read the material slowly and carefully and study the figures as you go along. Take frequent study breaks and review what you have already read before continuing further into the chapter.

b. Many of the concepts in this chapter relate to events in your life. It is always easier to remember a new idea by relating it to something you already know or have experienced. Look for these connections as you study.

Student Media

Activities

Energy Concepts
The Structure of ATP
How Enzymes Work
Membrane Structure
Diffusion
Facilitated Diffusion
Osmosis and Water Balance in Cells
Active Transport
Exocytosis and Endocytosis
Cell Signaling

Case Studies in the Process of Science

How Is the Rate of Enzyme Catalysis Measured?
How Does Osmosis Affect Cells?
How Do Cells Communicate with Each Other?

Basic Energy Concepts

Videos

Plasmolysis

Turgid *Elodea*

Organizing Tables

Compare the reactants, products, and efficiencies of cellular respiration and the use of gasoline in an engine by completing the table below.

TABLE 5.1			
	Reactants	Products	Efficiency (Percent of energy used to do work)
Cellular respiration			
Burning of gasoline in an automobile engine			

Compare the processes of passive transport, facilitated diffusion, and active transport.

TABLE 5.2			
	Passive Transport	Facilitated Diffusion	Active Transport
Does the process require the use of ATP?			
What special membrane proteins, if any, are needed?			
Describe an example in a cell.			

Compare the reactions of animal and plant cells when placed into isotonic, hypotonic, and hypertonic solutions. Compare your results to Figure 5.14 of your text.

TABLE 5.3

	Isotonic Solution	Hypotonic Solution	Hypertonic Solution
Animal cell			
Plant cell			

Content Quiz

Directions: Identify the *one* best answer for the multiple-choice questions. For true/false questions, determine if the statement is true or false. If false, change the underlined word(s) to make the statement true. Finally, add the correct word(s) to the fill-in-the-blank questions to make the statements true.

Biology and Society: Stonewashing Without the Stones

1. Which one of the following statements about cellulase is *false*? Cellulase:
   A. is a protein.
   B. breaks down cellular membranes.
   C. is a catalyst.
   D. is used by bacteria and fungi to break down plant material.
   E. is used in the process of biostoning.

2. True or False?   Using pumice stones is <u>more</u> friendly to the environment than using enzymes to produce stonewashed jeans.

3. The enzyme cellulase breaks down the polysaccharide _____.

Some Basic Energy Concepts
CONSERVATION OF ENERGY

4. You are riding a bike up and down hills. At which point do you have the greatest potential energy?
   A. at the bottom of the hill
   B. at the top of the hill
   C. riding down the hill at the fastest speed
   D. climbing the hill

5. True or False?   Energy is defined as the <u>capacity to do work</u>.

6. The principle known as _____ states that it is not possible to destroy or create energy.

ENTROPY

7. All energy conversions:
   A. destroy some energy.
   B. decrease the entropy of the universe.
   C. generate some heat.
   D. produce ATP.

8. True or False?   The heat produced by an automobile engine is a type of <u>kinetic energy</u>.

9. The term _____ is used as a measure of disorder or randomness.

CHEMICAL ENERGY

10. We feel warmer when we exercise because of extra heat produced by:
    A. cellular respiration.
    B. breathing faster.
    C. friction of blood flowing through the body.
    D. our movement through the air around us.
    E. sweating.

11. Examine Figure 5.3 of your text. Which one of the following is *not* produced by the engine and cellular respiration?
    A. heat
    B. water
    C. oxygen
    D. carbon dioxide

12. True or False?   The chemical energy in molecules of glucose and other fuels is a special type of <u>kinetic</u> energy.

13. Molecules of carbohydrates, fats, and gasoline all have structures that make them especially rich in _____ energy.

FOOD CALORIES

14. The energy in a 300-food-calorie cheeseburger could raise the temperature of:
    A. 1 kilogram (kg) of water by 30°C.
    B. 3 kg of water by 100°C.
    C. 3 kg of water by 1°C.
    D. 30 kg of water by 30°C.
    E. 30 kg of water by 100°C.

15. True or False?   The calories in food are a form of <u>potential</u> energy.

16. The amount of energy needed to raise 1 gram of water 1° Celsius is a(n) _____ while 1,000 times this amount of energy is a(n) _____.

ATP and Cellular Work
THE STRUCTURE OF ATP, PHOSPHATE TRANSFER

17. Moving ions across cell membranes is an example of what type of ATP work?
    A. chemical work
    B. transport work
    C. neutral work
    D. mechanical work

18. True or False?   ATP has <u>less</u> potential energy than ADP.

19. The energy for most cellular work is found in the _____ of the ATP molecule.

THE ATP CYCLE

20. What happens to the ADP molecule produced when ATP loses a phosphate during an energy transfer?
    A. ADP is used to build carbohydrates, fats, and proteins for use throughout the cell.
    B. ADP is released from the cells.
    C. ADP is broken down further into carbon atoms that are then reassembled into proteins.
    D. Energy from cellular respiration is used to convert ADP back to ATP.
    E. None of the above statements are correct.

21. Which one of the following processes is most like energy coupling?
    A. burning coal to produce electricity to run a factory
    B. running a furnace to heat a home
    C. using a car to drive to work
    D. using the energy from the sun to make sugar

22. During the process of energy coupling in a cell, _____ molecules are recycled.

Enzymes
ACTIVATION ENERGY

23. Which of the following can be used to initiate a chemical reaction?
    A. addition of ATP
    B. heat and the addition of ATP
    C. enzymes
    D. heat
    E. heat or enzymes

24. True or False? Enzymes <u>raise</u> the activation energy to break the bonds of reactant molecules.

25. The many chemical reactions that occur in organisms are collectively called _____.

INDUCED FIT

26. When a substrate molecule slips into an active site, the active site changes shape slightly to embrace the substrate and catalyze the reaction. This interaction is called:
    A. substrate locking.
    B. active engagement.
    C. active site shifting.
    D. induced fit.
    E. substrate embracing.

27. True or False? Most enzymes are named for their <u>inhibitors</u>.

28. The reactant molecule called the _____ binds at an enzyme's _____.

ENZYME INHIBITORS

29. Examine the role of an inhibitor in Figure 5.10 of your text. The inhibitor functions most like a:
    A. teacher giving you instructions for a laboratory exercise.
    B. friend who helps you do your laundry.
    C. person who has parked in your parking spot.
    D. screwdriver used to tighten screws.
    E. traffic light indicating to a driver when it is safe to cross an intersection.

30. True or False?   The inhibition of an enzyme by its product is an example of <u>feedback regulation</u>.

31. Many antibiotics that kill disease-causing bacteria are also enzyme _____.

Membrane Function
PASSIVE TRANSPORT: DIFFUSION ACROSS MEMBRANES

32. Releasing fish into a pond and seeing them swim in all directions is most like the process of:
    A. diffusion.
    B. osmosis.
    C. osmoregulation.
    D. active transport.
    E. pinocytosis.

33. Which one of the following most relies upon the process of diffusion?
    A. recognizing a friend eating at another table in a cafeteria
    B. discussing a subject in preparation for an exam
    C. selecting a new perfume or cologne
    D. listening to a song on the radio
    E. checking your pulse after exercise

34. True or False?   Cells <u>do not</u> need to spend energy for diffusion to occur.

35. A cell does not have to use _____ for passive transport to occur.

36. Some substances have properties that prevent them from directly diffusing across a membrane. Such substances, however, can diffuse across a membrane with the help of specific transport proteins in the process of _____.

OSMOSIS AND WATER BALANCE IN CELLS

37. If you soak your hands in dishwater, you may notice that your skin soaks up water and swells into distinct wrinkles. This is because your skin cells are _____ to the _____ dishwater.
    A. hypotonic . . . hypertonic
    B. hypertonic . . . hypotonic
    C. hypotonic . . . hypotonic
    D. isotonic . . . hypotonic
    E. hypertonic . . . isotonic

38. You decide to buy a new angelfish for your freshwater aquarium. When you introduce the fish into its new tank, the fish swells up and dies. You later learn it was an angelfish from the ocean. The unfortunate fish went from a _____ solution into a(n) _____ solution.

    A. mesotonic . . . hypotonic

    B. hypertonic . . . isotonic

    C. hypertonic . . . hypotonic

    D. hypotonic . . . isotonic

39. Examine Figure 5.14 of your text. Why does the animal cell but not the plant cell burst when it is in a hypotonic solution?

    A. The plant cell does not absorb as much salt.

    B. The plant cell is less hypertonic.

    C. The cell wall keeps the cell from bursting.

    D. The water cannot easily move through the plant cell wall.

40. True or False?   A crab in the ocean has the same salt concentration in its body as the surrounding seawater. Thus, the ocean is <u>hypotonic</u> to the crab.

41. When a *Paramecium* uses its contractile vacuole to remove excess water, it is engaged in the process called _____.

ACTIVE TRANSPORT: THE PUMPING OF MOLECULES ACROSS MEMBRANES

42. Moving a molecule across a membrane and against its concentration gradient requires:

    A. phospholipids using passive transport.

    B. phospholipids using active transport.

    C. membrane transport proteins using active transport.

    D. membrane transport proteins using passive transport.

    E. membrane transport proteins using receptor-mediated endocytosis.

43. True or False?   Membrane proteins using <u>passive</u> transport require ATP as a source of energy.

44. A nerve cell maintains a higher concentration of potassium ions inside itself than in its surroundings. This nerve cell is most likely using _____ transport to keep the potassium levels high.

EXOCYTOSIS AND ENDOCYTOSIS: TRAFFIC OF LARGE MOLECULES

45. Which one of the following is a type of endocytosis in which very specific molecules are brought into a cell using specific membrane protein receptors?
    A. osmoregulation
    B. phagocytosis
    C. pinocytosis
    D. receptor-mediated endocytosis
    E. none of the above

46. True or False?   Tears and saliva are products released by cells using endocytosis.

47. In _____, the cell gulps droplets of fluid by forming tiny vesicles.

THE ROLE OF MEMBRANES IN CELL SIGNALING

48. In a cell, the _____ relays a signal and converts it to chemical forms that work within the cell.
    A. process of exocytosis
    B. process of endocytosis
    C. process of pinocytosis
    D. receptor protein
    E. signal transduction pathway

49. True or False?   The three stages of cell signaling are reception, transduction, and response.

50. In a signal transduction pathway, external signals are received by specific receptor _____.

Evolution Connection: Evolving Enzymes

51. Which one of the following statements is *false*? Directed evolution:
    A. has a specific purpose chosen by the researchers.
    B. may require just a few weeks to produce a new enzyme.
    C. relies upon the natural environment.
    D. can be used to produce enzymes that perform a new function.

52. True or False?   Natural selection does not have purpose or direction.

53. Research data suggest that many of our human genes arose through _____ evolution.

Word Roots

endo = within, inner (endocytosis: taking material into a cell)

exo = outside (exocytosis: eliminating some materials outside of a cell)

hyper = excessive (hypertonic: in comparing two solutions, it refers to the one with the greater concentration of solutes)

hypo = lower (hypotonic: in comparing two solutions, it refers to the one with the lower concentration of solutes)

iso = same (isotonic: solutions with equal concentrations of solutes)

kinet = move (kinetic: type of energy, it is the energy of motion)

phago = eat (phagocytosis: cellular eating)

pino = drink (pinocytosis: cellular drinking)

tonus = tension (isotonic: solutions with equal concentrations of solutes)

Key Terms

activation energy
active site
active transport
ADP
ATP
calorie
cellular respiration
chemical energy
conservation of energy
diffusion

energy
energy coupling
entropy
enzyme
enzyme inhibitor
endocytosis
exocytosis
facilitated diffusion
feedback regulation

heat
hypertonic
hypotonic
induced fit
isotonic
kinetic energy
metabolism
osmoregulation
osmosis
passive transport
phagocytosis

pinocytosis
plasmolysis
potential energy
receptor-mediated endocytosis
signal transduction pathway
substrate
transport proteins

Crossword Puzzle

Use the Key Terms list from this chapter to fill in the crossword puzzle.

ACROSS

2. the type of endocytosis in which specific molecules move into a cell by inward budding

3. the interaction between a substrate molecule and the active site of an enzyme

5. a specific substance (reactant) on which an enzyme acts

6. cellular eating

8. the control of water and solute balance in an organism

9. having the same solute concentration as another solution

10. of two solutions, the one with the greater concentration of solutes

11. the energy of motion

12. the amount of energy associated with the movement of the atoms and molecules in a body of matter

13. protein that serves as a biological catalyst

14. the passage of a substance across a biological membrane down its concentration gradient

16. the type of energy that is stored

18. the transfer of energy from processes that yield energy to those that consume it

19. the many chemical reactions that occur in organisms

21. the movement of materials into the cytoplasm of a cell via membranous vesicles

23. the type of energy stored in the chemical bonds of molecules

24. type of pathway that converts a signal on a cell's surface to an inner response

26. the movement of materials out of the cytoplasm of a cell via membranous vesicles

28. diffusion across a membrane without the input of energy

29. the capacity to perform work

30. the type of energy that reactants must absorb before a chemical reaction will start

32. plant cell shriveling when there is a shortage of water

34. the part of an enzyme molecule where a substrate molecule attaches

35. the principle that energy can neither be created nor destroyed

DOWN

1. cellular drinking

4. a metabolic control in which the product inhibits the process that produced it

7. of two solutions, the one with the lesser concentration of solutes

13. a measure of disorder, or randomness

15. chemical that interferes with an enzyme's activities

17. the aerobic harvesting of energy from food molecules

20. the type of protein that helps move substances across a cell membrane

22. the tendency of molecules to move from high to low concentrations

25. the amount of energy that raises the temperature of 1 gram of water by 1° C

27. the passive transport of water across a selectively permeable membrane

30. a molecule composed of adenosine and two phosphate groups

31. the type of transport moving a substance across a membrane against its concentration gradient

33. a molecule composed of adenosine and three phosphate groups

CHAPTER 6

Cellular Respiration: Obtaining Energy from Food

Studying Advice

    a. This chapter addresses many abstract ideas that may initially be difficult to grasp. Do not try to understand all of this chapter in a single night. Read the material slowly and carefully and study the figures as you go along. Take frequent study breaks and review what you have already read before continuing further into the chapter.

    b. The organizing tables below should help you understand the basics of cellular respiration. Focus first on the overall process. Then read to gather the details. Table 6.1 should help you organize the definitions of the many pairs of contrasting terms.

Student Media

Activities

Build a Chemical Cycling System
Overview of Cellular Respiration
Glycolysis
The Citric Acid Cycle
Electron Transport
Fermentation

Case Studies in the Process of Science

How Is the Rate of Cellular Respiration Measured?

eTutors

Cellular Respiration (animation)

Organizing Tables

Compare the definitions of the following pairs of terms.

TABLE 6.1		
Photosynthesis	vs.	Respiration
Autotrophs	vs.	Heterotrophs
Producers	vs.	Consumers
Aerobic	vs.	Anaerobic
Facultative anaerobe	vs.	Obligate anaerobe
Obligate aerobe	vs.	Obligate anaerobe

Compare the processes of glycolysis, citric acid cycle, and electron transport chain in the table below. Some cells are already filled in to make the job a little easier!

TABLE 6.2				
	Location	Reactants	Products	Energy Yield
Glycolysis			Two molecules of pyruvic acid	
Citric acid cycle		Acetyl CoA	CO_2	
Electron transport chain		NADH		

Compare the products of each of the following reactions.

TABLE 6.3

Process	Products
Fermentation in human muscle cells	
Fermentation in yeast	

Content Quiz

Directions: Identify the *one* best answer for the multiple-choice questions. For true/false questions, determine if the statement is true or false. If false, change the underlined word(s) to make the statement true. Finally, add the correct word(s) to the fill-in-the-blank questions to make the statements true.

Biology and Society: Feeling the Burn

1. Which one of the following does *not* occur as you approach your aerobic capacity?
   A. You use aerobic metabolism.
   B. Oxygen is used to generate ATP.
   C. Your muscle cells use increasing amounts of oxygen.
   D. Your muscle cells produce lactic acid.

2. True or False?   Lactic acid is produced during <u>aerobic</u> metabolism.

3. The burning sensation of an overexercised muscle results from the buildup of _____.

Energy Flow and Chemical Cycling in the Biosphere
PRODUCERS AND CONSUMERS

4. Which one of the following statements about photosynthesis is *false*? Photosynthesis:
   A. occurs mainly within the roots of plants.
   B. provides the food for most ecosystems.
   C. provides the energy to power chemical processes that make organic molecules.
   D. occurs within chloroplasts.

5. True or False? <u>Chlorophyll</u> in chloroplasts makes plant leaves appear green.

6. True or False? Plants and other photosynthetic organisms are the <u>consumers</u> in an ecosystem.

7. Organisms that *cannot* make organic molecules from inorganic ones are called _____.

CHEMICAL CYCLING BETWEEN PHOTOSYNTHESIS AND CELLULAR RESPIRATION

8. Plants use photosynthesis to join together:
    A. carbon dioxide and water to make oxygen and glucose.
    B. carbon dioxide and glucose to make water and oxygen.
    C. carbon dioxide and oxygen to make water and glucose.
    D. glucose and oxygen to make water and carbon dioxide.
    E. water and oxygen to make glucose and carbon dioxide.

9. Cellular respiration:
    A. occurs in chloroplasts.
    B. is used by plants, but not animals.
    C. is part of photosynthesis.
    D. harvests energy stored in sugars and other organic molecules.
    E. produces glucose and oxygen.

10. Examine Figure 6.3 of your text. In this ecosystem, the carbon dioxide that the trees use in photosynthesis came from:
    A. plants.
    B. wolves.
    C. rabbits.
    D. all of the above.
    E. none of the above.

11. True or False? Cellular respiration occurs in plant and animal cells in organelles called <u>chloroplasts</u>.

12. Cellular respiration typically uses energy extracted from organic fuel to produce another form of chemical energy called _____.

13. Photosynthesis combines together _____ and _____, which are products of cellular respiration.

Cellular Respiration: Aerobic Harvest of Food Energy
THE RELATIONSHIP BETWEEN CELLULAR RESPIRATION AND BREATHING

14. What do cellular respiration and breathing have in common? Both processes:
    A. produce ATP.
    B. produce glucose.
    C. take in carbon dioxide and release oxygen.
    D. take in oxygen and release carbon dioxide.

15. True or False? An aerobic cellular process requires <u>carbon dioxide</u>.

16. The aerobic harvesting of chemical energy from organic food molecules defines _____.

THE OVERALL EQUATION FOR CELLULAR RESPIRATION

17. Which one of the following is *not* produced as a result of the breakdown of a single glucose molecule by cellular respiration?
    A. 6 molecules of carbon dioxide
    B. 6 molecules of water
    C. 6 molecules of oxygen
    D. up to 38 ATP molecules

18. True or False? Cellular respiration consists of <u>many</u> steps.

19. Oxygen is a vital part of aerobic respiration because it accepts _____ atoms from glucose.

THE ROLE OF OXYGEN IN CELLULAR RESPIRATION

20. Which one of the following occurs during the transfer of hydrogen in cellular respiration?
    A. Glucose is reduced, oxygen is oxidized, and energy is released.
    B. Glucose is oxidized, oxygen is reduced, and energy is released.
    C. Glucose is reduced, oxygen is oxidized, and energy is consumed.
    D. Glucose is oxidized, oxygen is reduced, and energy is consumed.

21. Which one of the following does *not* occur during the cellular respiration of glucose?
    A. NAD^+ donates electrons to NADH.
    B. NADH transfers electrons from glucose to the top of the electron transport chain.
    C. Electrons cascade down the electron transport chain giving up a small amount of energy with each transfer.
    D. Oxygen is the final electron acceptor at the bottom of the electron transport chain.

22. Examine Figure 6.6 of your text. The transfer of electrons during cellular respiration is most like:

    A. a giant waterfall.

    B. walking down a set of stairs.

    C. climbing a mountain.

    D. running up a set of stairs.

    E. playing catch with a baseball.

23. True or False?   In the electron transport chain, NAD^+ functions like gravity, pulling electrons down the chain.

24. NADH undergoes the process of _____ when it donates its electrons to the electron transport chain.

THE METABOLIC PATHWAY OF CELLULAR RESPIRATION

25. Which one of the following statements about cellular respiration is *false*?

    A. ATP synthase, located within the inner mitochondrial membrane, helps form ATP.

    B. Cellular respiration is a metabolic pathway consisting of more than two dozen chemical reactions, each catalyzed by a specific enzyme.

    C. Inside mitochondria, glycolysis joins a pair of three-carbon pyruvic acid molecules to form a molecule of glucose.

    D. The citric acid cycle breaks down molecules of acetyl-CoA to release CO_2 and energy trapped by NADH.

26. The extensive infolding of the inner mitochondrial membrane is likely an adaptation to:

    A. make it more difficult for oxygen to pass through.

    B. increase the flexibility of mitochondria.

    C. make it more difficult for glucose to pass through.

    D. increase the surface area of the electron transport system.

    E. increase the surface area for glycolysis.

27. Examine Figure 6.11 of your text. What happens to the two carbons in Acetyl CoA during the citric acid cycle? The two carbons are:

    A. attached to NADH.

    B. used to make glucose.

    C. released as CO_2.

    D. added to ADP to make ATP.

    E. destroyed in the process.

28. Most ATP is generated during the process of:
    A. glycolysis.
    B. fermentation.
    C. electron transport.
    D. citric acid cycle.

29. Energy from the electron transport chain pumps _____ across the inner mitochondrial membrane.

30. Cellular respiration is an example of _____ , the sum of all of the chemical processes that occur in cells.

Fermentation: Anaerobic Harvest of Food Energy
FERMENTATION IN HUMAN MUSCLE CELLS

31. Which one of the following statements about fermentation is *false*?
    A. Your cells can produce ATP when no oxygen is present.
    B. When your muscles use fermentation, glucose is produced.
    C. Fermentation in your cells uses the process of glycolysis.
    D. Fermentation produces about two ATP per glucose molecule.
    E. Fermentation occurs when oxygen levels are not sufficient to support cellular respiration.

32. True or False?   Fermentation is an <u>anaerobic</u> process.

33. During the process of fermentation, pyruvic acid is converted to _____.

FERMENTATION IN MICROORGANISMS

34. Yeast used to make beer and bread use fermentation to produce:
    A. ATP, carbon dioxide, and oxygen.
    B. ATP, water, and oxygen.
    C. ATP, ethyl alcohol, and carbon dioxide.
    D. ethyl alcohol and oxygen.
    E. carbon dioxide and water.

35. True or False?   Although our cells behave as <u>facultative anaerobes</u>, our entire body is an <u>obligate aerobe</u>.

36. Some organisms, called _____ anaerobes, are unable to live in the presence of oxygen.

Matching: For each of the following, indicate whether it is a facultative anaerobe, obligate anaerobe, or obligate aerobe.

_____ 37. yeast

_____ 38. human muscle cells

_____ 39. an entire human organism

_____ 40. bacteria living in stagnant ponds or deep soil

A. facultative anaerobe

B. obligate anaerobe

C. obligate aerobe

Evolution Connection: Life on an Anaerobic Earth

41. Glycolysis is considered to be an ancient metabolic process because:
    A. oxygen was abundant in the early Earth atmosphere.
    B. glycolysis is found only in animals.
    C. glycolysis occurs in the cytosol.
    D. it produces more ATP per glucose molecule than cellular respiration.

42. True or False? The process of glycolysis may be <u>more than</u> 3 billion years old.

43. The early Earth atmosphere had very little _____, favoring anaerobic organisms that could use glycolysis.

Word Roots

aero = air (aerobic: chemical reaction using oxygen)

auto = self; **troph** = food (autotroph: organism that makes its own organic matter)

glyco = sweet; **lysis** = split (glycolysis: process that splits glucose into two molecules)

hetero = other (heterotroph: Greek word that means "other feeder")

oblig = bound (obligate aerobe: organism that must have oxygen)

photo = light (photosynthesis: process using light energy to make organic molecules)

Key Terms

aerobic	citric acid cycle	fermentation	oxidation
aerobic capacity	consumer	glycolysis	photosynthesis
anaerobic	electron transport	heterotroph	producer
ATP synthase	chain	NADH	redox reaction
autotroph	facultative	obligate aerobe	reduction
cellular respiration	anaerobe	obligate anaerobe	

Crossword Puzzle

Use the Key Terms list from this chapter to fill in the crossword puzzle.

1. type of organism that uses photosynthesis
3. abbreviation for type of reaction that transfers electrons
4. process using light energy to make organic molecules
6. type of chemical reaction that does not use oxygen
7. organism that cannot make its own organic food molecules
9. chemical reaction using oxygen
11. organism that makes all of its own organic matter
12. a type of anaerobe that can make ATP using aerobic or anaerobic respiration
14. cluster of proteins built into the inner mitochondrial membrane
16. a molecule that carries electrons from glucose to the electron transport chain
17. process that harvests energy stored in sugars
18. anaerobic harvest of food energy
20. type of cycle that continues breakdown of glucose after glycolysis

2. loss of electrons during a redox reaction
5. the maximum rate that oxygen can be taken in and used by muscle cells
8. a type of chain composed of electron carrier molecules used to make ATP
10. a type of anaerobe that cannot survive around oxygen
13. heterotroph that eats other heterotrophs or autotrophs
15. process that splits glucose into two molecules
19. acceptance of electrons during a redox reaction

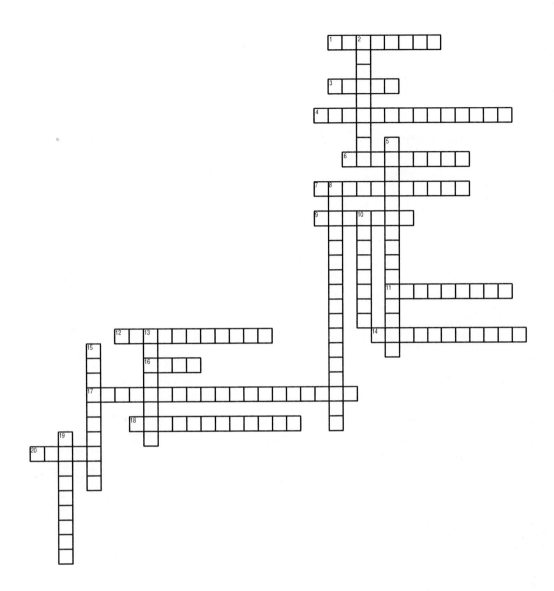

CHAPTER 7

Photosynthesis: Using Light to Make Food

Studying Advice

a. Like chapters 5 and 6, this chapter addresses many abstract ideas that may initially be difficult to grasp. Do not try to understand this entire chapter in a single night. Read the material slowly and carefully and study the figures as you go along. Figures 7.10–7.13 are especially helpful in understanding the molecular details of photosynthesis. Take frequent study breaks and review what you have already read before continuing further into the chapter.

b. The organizing table below should help you understand the two basic steps of photosynthesis and compare these to cellular respiration.

Student Media

Activities

The Plants in Our Lives

The Sites of Photosynthesis

Overview of Photosynthesis

Light Energy and Pigments

The Light Reactions

The Calvin Cycle

Photosynthesis in Dry Climates

The Greenhouse Effect

Case Studies in the Process of Science

How Does Paper Chromatography Separate Plant Pigments?
How Is the Rate of Photosynthesis Measured?

eTutors

Photosynthesis

MP3 Tutors

Photosynthesis

Videos

Discovery Channel Video Clip: Space Plants

Organizing Tables

Compare the reactants, products, and location in the cell for the three reactions listed at the left.

TABLE 7.1			
	Reactants	Products	Location in the Cell
Cellular respiration			
Light reaction			
Calvin cycle			

Content Quiz

Directions: Identify the *one* best answer for the multiple-choice questions. For true/false questions, determine if the statement is true or false. If false, change the underlined word(s) to make the statement true. Finally, add the correct word(s) to the fill-in-the-blank questions to make the statements true.

Biology and Society: Plant Power for Power Plants

1. During most of human history, what has been the main source of energy for heat, light, and fuel for cooking?
   A. wood
   B. oil
   C. coal
   D. natural gas
   E. nuclear

2. Burning wood for fuel has many advantages over using fossil fuels. Which one of the following is *not* an advantage of burning wood instead of fossil fuels? Energy plantations:
   A. provide wildlife habitat.
   B. reduce soil erosion.
   C. renew the soil.
   D. increase carbon dioxide in the air.

3. True or False?   All of the food consumed by humans can be traced back to animals.

4. Over the last century, wood has been largely displaced as an energy source by _____.

The Basics of Photosynthesis
CHLOROPLASTS: SITES OF PHOTOSYNTHESIS

5. Photosynthesis commonly occurs when sunlight hits chlorophyll pigments in the:
   A. thylakoid membrane of a mitochondrion in a leaf.
   B. stroma of a chloroplast in a cell of a leaf.
   C. thylakoid membrane of a chloroplast in a cell surrounding a stoma of a leaf.
   D. thylakoid membrane of a chloroplast in a cell of a leaf.

6. Examine Figure 7.3 to compare the drawing and photograph of a chloroplast. How is the drawing different from the photograph of an actual chloroplast from a plant leaf?

   A. The grana are not stacked in the drawing.

   B. The grana are not interconnected in the drawing.

   C. There is no stroma in the drawing.

   D. The intermembrane space is much larger in the drawing.

7. True or False?   Sugars are made in the grana of a chloroplast.

8. Carbon dioxide enters the leaf, and oxygen exits, through tiny pores called _____.

THE OVERALL EQUATION FOR PHOTOSYNTHESIS

9. Which one of the following statements about photosynthesis is *false*? Photosynthesis:

   A. uses the products of respiration.

   B. produces carbon dioxide and oxygen.

   C. uses energy from sunlight to split water and release oxygen into the atmosphere.

   D. boosts the energy in electrons "uphill" to join carbon dioxide and hydrogens.

10. True or False?   Inside chloroplasts, carbon dioxide is split to form hydrogen and oxygen.

11. The largest molecule produced by photosynthesis is _____.

A PHOTOSYNTHESIS ROAD MAP

12. ATP and NADPH, produced by the:

   A. light reaction, are used to reduce carbon dioxide to form glucose.

   B. light reaction, are used to reduce glucose to form carbon dioxide.

   C. Calvin cycle, are used to reduce carbon dioxide to form glucose.

   D. Calvin cycle, are used to reduce glucose to form carbon dioxide.

13. Examine the overall equation for photosynthesis. Which one of the following is a product of the Calvin cycle?

   A. ATP

   B. NADPH

   C. G3P

   D. carbon dioxide

   E. water

14. True or False?   The light reaction <u>does not</u> produce sugar.

15. The Calvin cycle depends on a supply of _____ and _____ from the light reactions.

The Light Reactions: Converting Solar Energy to Chemical Energy
THE NATURE OF SUNLIGHT

16. Which one of the following statements about radiation is *false*?
    A. Waves in the electromagnetic spectrum carry energy.
    B. The electromagnetic spectrum is the full range of radiation.
    C. Energy in wavelengths of visible light is absorbed by components of the Calvin cycle.
    D. The wavelengths of visible light that we see were not absorbed by what we are looking at.
    E. A wavelength is the distance between the crests of two adjacent waves.

17. True or False?   We can see <u>only a small portion of</u> the entire electromagnetic spectrum.

18. A dark blue shirt will absorb _____ light energy than a very light blue shirt.

THE PROCESS OF SCIENCE: WHAT COLORS OF LIGHT DRIVE PHOTOSYNTHESIS?

19. What did Engelmann use to determine where the highest concentrations of oxygen were produced?
    A. oxygen-sensitive bacteria
    B. an electronic oxygen probe
    C. signs of bubbles of oxygen gas accumulating
    D. the speed of particles moving away from the oxygen-producing algae
    E. a change in the color of the water around the oxygen-producing algae

20. True or False?   Chloroplasts use <u>all</u> of the wavelengths of light to drive photosynthesis.

21. Most plants least use the wavelengths of light that form the color _____.

CHLOROPLAST PIGMENTS

22. Chlorophyll *b*:
    A. absorbs mainly blue-violet and red light.
    B. participates directly in the light reactions.
    C. absorbs and dissipates excessive light that could damage chlorophyll.
    D. participates indirectly in the light reactions by transferring light energy to chlorophyll *a*.

23. True or False?   Chloroplast pigments are all located within the <u>outer</u> membrane of the chloroplast.

24. Only the pigment _____ participates directly in the light reactions.

HOW PHOTOSYSTEMS HARVEST LIGHT ENERGY

25. The photosynthetic pigments work together to function most like:
    A. a telephone.
    B. a brain.
    C. a mirror.
    D. an umbrella.
    E. an antenna.

26. Which one of the following is the general sequence of energy transfer during photosynthesis? Light energy in a photon is transferred first to:
    A. a primary electron acceptor, then to chlorophyll *a*, and finally to pigment molecules.
    B. pigment molecules, then to a primary electron acceptor, and finally to chlorophyll *a*.
    C. pigment molecules, then to chlorophyll *b*, and finally to a primary electron acceptor.
    D. pigment molecules, then to chlorophyll *a*, and finally to a primary electron acceptor.
    E. none of the above.

27. True or False?   The reaction center of a photosystem consists of a chlorophyll *a* molecule next to a <u>primary electron acceptor</u>.

28. A fixed quantity of light energy is a(n) _____.

HOW THE LIGHT REACTIONS GENERATE ATP AND NADPH

29. Which one of the following parts of a mitochondrion functions most like the thylakoid membrane?
    A. mitochondrial matrix
    B. outer mitochondrial membrane
    C. inner mitochondrial membrane
    D. the intermembrane space

30. The extensive infolding of the inner thylakoid membrane is likely an adaptation to:
    A. make it more difficult for oxygen to pass through.
    B. increase the flexibility of chloroplasts.
    C. make it more difficult for hydrogen ions to pass through.
    D. increase the surface area of the light reactions.
    E. increase the surface area for glycolysis.

31. Oxygen is produced during the:
    A. water-splitting photosystem.
    B. electron transport chain.
    C. NADPH-producing photosystem.
    D. Calvin cycle.

32. True or False?   Both cellular respiration and photosynthesis have ATP synthases that use the energy stored by the H^+ gradient.

33. The electron transport chain in the chloroplast releases energy used to produce _____.

The Calvin Cycle: Making Sugar from Carbon Dioxide
WATER-SAVING ADAPTATIONS OF C_4 AND CAM PLANTS

34. Which of the following are produced by the Calvin cycle?
    A. $NADP^+$, ADP + P, and G3P
    B. NADPH, ATP, and glucose
    C. $NADP^+$, ADP + P, and glucose
    D. NADPH, ATP, and G3P
    E. glucose and oxygen

35. Which of the following *only* includes CAM plants?
    A. pineapples, cacti, and aloe
    B. soybeans, wheat, and jade
    C. soybeans, wheat, and cacti
    D. oats, wheat, and pineapples

36. True or False?   Plant cells use <u>G3P</u> to make glucose and other organic molecules they need.

37. CAM plants are most likely to have their _____ open mainly at night.

38. During a hot and dry summer, C_3 plants can lose dangerous amounts of _____ if their stomata remain open during the day.

The Environmental Impact of Photosynthesis
HOW PHOTOSYNTHESIS MODERATES GLOBAL WARMING

39. Deforestation contributes to global warming by:
    A. decreasing the number of producers and adding CO_2 to the atmosphere.
    B. decreasing the number of consumers and adding CO_2 to the atmosphere.
    C. increasing the number of producers and removing CO_2 from the atmosphere.
    D. increasing the number of consumers and removing CO_2 from the atmosphere.

40. True or False?   Increasing the total amount of photosynthesis on Earth would <u>increase</u> global warming by removing CO_2 from the atmosphere.

41. The _____ is a gradual warming of Earth caused by increased atmospheric carbon dioxide.

Evolution Connection: The Oxygen Revolution

42. Large amounts of oxygen were first added to Earth's atmosphere by:
    A. land plants.
    B. algae in the oceans.
    C. cyanobacteria.
    D. fungi.
    E. the greenhouse effect.

43. True or False?   The widespread production of oxygen by cyanobacteria 2.7–3.5 billion years ago permitted the evolution of <u>photosynthetic</u> organisms.

44. Many anaerobic prokaryotes die if they are exposed to _____.

Word Roots

chloro = green; **plast** = formed or molded (chloroplast: the organelle of photosynthesis)

electro = electricity; **magne** = magnetic (electromagnetic spectrum: the full range of radiation)

photo = light (photosystem: cluster of pigment molecules)

phyll = leaf (chlorophyll: photosynthetic pigment in chloroplasts)

stoma = mouth (stomata: tiny pores in leaves through which gases are exchanged)

thylac = a sac or pouch (thylakoids: membranous sacs suspended in the stroma)

Key Terms

C_3 plants	electromagnetic	light reaction	reaction center
C_4 plants	spectrum	NADPH	stomata
Calvin cycle	global warming	photon	stroma
CAM plants	grana	photosystem	thylakoids
chlorophyll *a*	greenhouse effect	primary electron	wavelength
chloroplast	greenhouse gases	acceptor	

Crossword Puzzle

Use the Key Terms list from this chapter to fill in the crossword puzzle.

ACROSS

1. a type of electron acceptor that harvests light
2. in a chloroplast, the chlorophyll *a* molecule and primary electron acceptor
4. tiny pores in leaves through which gases are exchanged
5. thick fluid deep inside a chloroplast
6. the gases in the atmosphere that absorb heat radiation
8. pigment that absorbs blue-violet and red light
9. the type of spectrum of all types of radiation
12. stacks of hollow disks inside chloroplasts
16. membranous sacs suspended in the stroma
17. changes in the surface air temperature induced by emission of greenhouse gases
18. the process of atmospheric carbon dioxide trapping heat
19. a fixed quantity of light energy

DOWN

1. cluster of pigment molecules
3. process that makes sugar from carbon dioxide
7. the organelle of photosynthesis
8. the type of plant that uses crassulacean acid metabolism
10. type of plant that prefaces the Calvin cycle with reactions that incorporate carbon dioxide into four-carbon compounds
11. type of plant that uses the Calvin cycle to form three-carbon compounds as the first stable intermediate
13. distance between crests of adjacent light waves
14. the type of reaction converting solar energy to chemical energy
15. electron carrier involved in photosynthesis

CHAPTER 8

Cellular Reproduction: Cells from Cells

Studying Advice

a. Read carefully and study Figures 8.8 and 8.16 of your text describing the key events of mitosis and meiosis. Figure 8.17 is a very useful comparison of both processes.

b. The key to understanding mitosis and meiosis is learning how the chromosomes are arranged and how they separate during metaphase. Again, Figures 8.8 and 8.16 of your text are crucial.

c. The organizing tables below should help you sort out the important details of the cell cycle, mitosis, and meiosis.

Student Media

Activities

Asexual and Sexual Reproduction

The Cell Cycle

Mitosis and Cytokinesis Animation

Mitosis and Cytokinesis Video

Causes of Cancer

Human Life Cycle

Meiosis Animation

The Origins of Genetic Variation

Polyploid Plants

Case Studies in the Process of Science

How Much Time Do Cells Spend in Each Phase of Mitosis?
How Can the Frequency of Crossing Over Be Estimated?

eTutors

Mitosis
Meiosis

MP3 Tutors

Mitosis
Meiosis
Mitosis-Meiosis Comparison

Videos

Animal Mitosis (time-lapse)
Hydra Budding

Organizing Tables

Compare the key events of each stage of the cell cycle.

TABLE 8.1	
	Key Events
M phase	
G$_1$ phase	
G$_2$ phase	
S phase	

Compare the following aspects of the G$_2$ phase of interphase and the stages of mitosis.

TABLE 8.2			
Stages	The Arrangement and Behavior of the Chromosomes	The Structure and Functions of the Mitotic Spindle	Other Cellular Details
G$_2$ Interphase: Preparing for mitosis			
Prophase			
Metaphase			
Anaphase			
Telophase			

Compare the following aspects of meiosis I and meiosis II.

TABLE 8.3			
Stage of Meiosis	How Are the Chromosomes Arranged During Metaphase? (Draw the arrangement of two tetrads)	What Separates during Anaphase: Homologous Chromosomes or Sister Chromatids?	Where Does Crossing Over Begin?
Meiosis I			
Meiosis II			

Content Quiz

Directions: Identify the *one* best answer for the multiple-choice questions. For true/false questions, determine if the statement is true or false. If false, change the underlined word(s) to make the statement true. Finally, add the correct word(s) to the fill-in-the-blank questions to make the statements true.

Biology and Society: A $50,000 Egg!

1. Which one of the following does not occur in the process of *in vitro* fertilization (IVF)?
   A. cloning sperm cells in a petri dish
   B. joining sperm and egg in a petri dish
   C. allowing an embryo to grow to the 8-cell stage
   D. implanting the embryo into the uterus of the mother

2. True or False?   Infertility affects one in <u>a thousand</u> American couples.

3. The inability to produce children after one year of trying defines _____.

What Cell Reproduction Accomplishes
PASSING ON GENES FROM CELL TO CELL

4. Daughter cells formed during typical cell division have:
   A. identical sets of chromosomes with identical genes.
   B. different sets of chromosomes with identical genes.
   C. identical sets of chromosomes with different genes.
   D. different sets of chromosomes with different genes.

5. True or False?   Growth of an organism and the replacement of lost or damaged cells are the main roles of <u>cell division</u>.

6. Before a cell divides, it duplicates its _____.

THE REPRODUCTION OF ORGANISMS

7. Sexual reproduction:
   A. uses only meiosis.
   B. involves the production of daughter cells with double the genetic material of the parent cell.
   C. uses only ordinary cell division.
   D. requires the fertilization of an egg by a sperm.
   E. is the normal process by which the cells of most organisms divide.

8. True or False?   A sperm or egg has <u>twice</u> as many chromosomes as its parent cell.

9. In _____ reproduction, the offspring and parent have identical genes.

The Cell Cycle and Mitosis

EUKARYOTIC CHROMOSOMES

10. Which one of the following statements about chromosomes is *false*?
    A. Sister chromatids remain attached until anaphase.
    B. The number of chromosomes in a eukaryotic cell depends upon the species.
    C. Chromosomes are made of a combination of DNA, carbohydrate, and lipid molecules.
    D. Human body cells each typically have 46 chromosomes.
    E. Once separated, sister chromatids go to different cells.

11. True or False?   The <u>lipids</u> in chromosomes help control the activity of the genes.

12. Sister chromatids remain attached at a region called the _____.

THE CELL CYCLE

13. What stage of the cell cycle is marked by the presence of sister chromatids and the preparation of the cell for division?
    A. S
    B. M
    C. G_1
    D. G_2
    E. G_3

14. True or False?   Some highly specialized cells <u>do not</u> undergo a cell cycle.

15. True or False?   Prokaryotes <u>do not</u> undergo mitosis.

16. The M phase of the cell cycle includes _____ and _____.

17. During the _____ stage of the cell cycle, the chromosomes are duplicated.

MITOSIS AND CYTOKINESIS

18. Which one of the following statements about mitosis is *false*?
    A. During prophase, the sister chromatids are clearly seen, the nuclear envelope breaks up, and the mitotic spindle begins to form.
    B. Toward the end of prophase, the sister chromatids replicate again, but remain attached to each other.
    C. During metaphase, the chromosomes line up in the middle of the mitotic spindle.
    D. At the start of anaphase, the sister chromatids are pulled apart and start to move toward opposite spindle poles.
    E. Telophase is like the opposite of prophase: The nuclear envelope reforms, the chromosomes uncoil, and the spindle disappears.

19. Dividing an animal cell into two cells is most like:
    A. building a new wall that divides one room into two.
    B. overtightening the drawstring of sweatpants.
    C. peeling an apple.
    D. cutting a pie into two pieces with a knife.

20. Examine the prophase stages of mitosis in Figure 8.8 of your text. Which of the following structures invade(s) the nuclear space after the nuclear membrane disintegrates?
    A. chromosomes
    B. centrioles
    C. nucleolus
    D. spindle microtubules

21. Compare the cell stages in Figure 8.8 of the text to the photograph in Figure 8.4. What stage is indicated by the cell in Figure 8.4?
    A. interphase
    B. early prophase
    C. late prophase
    D. metaphase
    E. anaphase

22. Spindle microtubules grow from _____, clouds of cytoplasmic material that in animal cells contains centrioles.

CANCER CELLS: GROWING OUT OF CONTROL

23. The most dangerous property of cancer cells is that they:
    A. divide slowly.
    B. spread throughout the body.
    C. form tumors.
    D. do not divide by mitosis.
    E. do not live very long.

24. Which one of the following can *reduce* your risk of developing cancer?
    A. smoking
    B. overexposure to the sun
    C. exercising
    D. Eating a low-fiber high-fat diet.

25. True or False?   Radiation and chemotherapy both fight cancer by disrupting cell division.

26. An abnormal mass of cells that remains at its original site is a _____ tumor.

27. The use of drugs to disrupt division of cancerous cells is called _____.

Meiosis, the Basis of Sexual Reproduction
HOMOLOGOUS CHROMOSOMES

28. A male human somatic cell has:
    A. 46 pairs of chromosomes.
    B. 23 pairs of sex chromosomes.
    C. 23 pairs of autosomes.
    D. 22 pairs of autosomes and 1 pair of sex chromosomes.
    E. 22 pairs of sex chromosomes and 1 pair of autosomes.

29. True or False?   Most of the chromosomes in humans are <u>sex chromosomes</u>.

30. The same sequence of genes is found on _____ chromosomes.

GAMETES AND THE LIFE CYCLE OF A SEXUAL ORGANISM

31. In the human life cycle somatic cells are:
    A. diploid, and gametes are diploid.
    B. diploid, and gametes are haploid.
    C. haploid, and gametes are diploid.
    D. haploid, and gametes are haploid.

32. True or False?   If the human reproductive cycle only included diploid cells, the offspring of each generation would have <u>twice as much</u> genetic material in each cell as the parent cells.

33. If the somatic cells of a species have 25 pairs of homologous chromosomes, the number of chromosomes in haploid gametes would be _____.

THE PROCESS OF MEIOSIS

34. Which one of the following statements about meiosis is *false*?
    A. Cells dividing by meiosis undergo two consecutive divisions.
    B. Homologous chromosomes exchange segments before separating from each other.
    C. Homologous chromosomes separate during meiosis I.
    D. Sister chromatids separate during meiosis II.
    E. The chromosomes replicate before meiosis and between meiosis I and meiosis II.

35. In sexually reproducing organisms, cells entering mitosis have:
    A. half the amount of DNA as cells entering meiosis.
    B. the same amount of DNA as cells entering meiosis.
    C. twice the amount of DNA as cells entering meiosis.
    D. four times the amount of DNA as cells entering meiosis.

36. True or False?   Before crossing over occurs, sister chromatids are <u>identical</u>.

37. The result of meiosis is four _____ cells.

REVIEW: COMPARING MITOSIS AND MEIOSIS

38. Which one of the following is an actual difference between mitosis and meiosis?
    A. A single cell is divided into two cells in mitosis and four cells in meiosis.
    B. Mitosis produces haploid cells, and meiosis produces diploid cells.
    C. Mitosis involves two cellular divisions, and meiosis involves just one cellular division.
    D. The chromosomes replicate before mitosis and meiosis, but in meiosis they replicate again between the first and second division.

39. True or False?   The two daughter cells produced by meiosis I <u>do not</u> undergo cytokinesis before meiosis II.

40. Sister chromatids are separated during mitosis and _____.

THE ORIGINS OF GENETIC VARIATION

41. The random segregation of one member of each homologous pair of chromosomes into gametes defines:
    A. random fertilization.
    B. crossing over.
    C. independent assortment.
    D. random assortment.
    E. independent fertilization.

42. True or False?   Crossing over only occurs in <u>mitosis</u>.

43. The site of crossing over is a called a(n) _____.

WHEN MEIOSIS GOES AWRY

44. In general:
    A. the absence of a Y chromosome results in maleness.
    B. the presence of only a single X chromosome results in maleness.
    C. a human embryo with an abnormal number of chromosomes is usually normal.
    D. nondisjunction occurs only in males of sexually reproducing, diploid organisms.
    E. the incidence of Down syndrome increases markedly with the age of the mother.

45. Examine the chart in Figure 8.23 of your text. At which of the following intervals do we find the greatest *increase* in risk of having a child with Down syndrome?

   A. 25–30 years

   B. 30–35 years

   C. 35–40 years

   D. 40–45 years

   E. 45–50 years

46. True or False?   Klinefelter syndrome and Turner syndrome result from nondisjunction of <u>autosomes</u>.

47. A person with trisomy 21 is said to have _____.

Evolution Connection: New Species from Errors in Cell Division

48. Polyploid species:

   A. are common in mammals.

   B. are more common in animals than in plants.

   C. can result if gametes are produced by mitosis.

   D. have two sets of chromosomes in each somatic cell.

49. True or False?   At least half of all species of flowering plants are <u>polyploid</u>.

50. The union of a diploid egg and a diploid sperm will produce a _____ zygote.

Word Roots

a = not or without (asexual: type of reproduction not involving fertilization)

ana = again (anaphase: mitotic stage when sister chromatids separate)

auto = self (autosomes: the chromosomes that do not determine gender)

carcin = an ulcer (carcinoma: cancer originating in the external or internal coverings of the body)

centro = the center; **mere** = a part (centromere: the centralized region joining two sister chromatids)

chemo = chemical (chemotherapy: type of cancer therapy using drugs that disrupt cell division)

chiasm = cross-mark (chiasma: the sites where crossing over have occurred)

chroma = colored (chromosome: DNA-containing structure)

cyto = cell; **kinet** = move (cytokinesis: division of the cytoplasm)

di = two (diploid: cells that contain two homologous sets of chromosomes)

fertil = fruitful (fertilization: process of fusion of sperm and egg cell)

gamet = a wife or husband (gamete: egg or sperm)

haplo = single (haploid: cells that contain only one chromosome of each homologous pair)

homo = like (homologous: like chromosomes that form a pair)

inter = between (interphase: time when a cell metabolizes and performs its various functions)

karyo = nucleus (karyotype: a display of the chromosomes of a cell)

leuko = white (leukemia: cancer of white blood cells)

mal = bad or evil (malignant: type of tumor that migrates away from its site of origin)

mei = less (meiosis: the division of a diploid nucleus into four haploid daughter nuclei)

meta = between (metaphase: mitotic stage when the chromosomes are lined up in the cell's middle)

mito = a thread (mitosis: the division of a diploid cell into two diploid cells)

non = not; **dis** = separate (nondisjunction: the result when paired chromosomes fail to separate)

poly = many (polyploid: having more than two sets of homologous chromosomes in each somatic cell)

pro = before (prophase: mitotic stage when the nuclear membrane first breaks up)

sarco = flesh (sarcoma: cancer that arises in tissues that support the body)

soma = body (somatic: body cells with 46 chromosomes in humans)

telo = end (telophase: final mitotic stage when the nuclear envelope reforms)

tetra = four (tetrad: the orientation of homologous pairs of chromosomes with four apparent "arms")

tri = three (trisomy 21: a condition in which a person has three number 21 chromosomes)

Key Terms

anaphase
asexual
  reproduction
autosome
benign tumor
cancer
carcinomas
cell cycle
cell cycle control
  system
cell division
cell plate
centromere
centrosome
chemotherapy

chiasma
chromatin
chromosome
cleavage furrow
crossing over
cytokinesis
diploid
Down syndrome
fertilization
gamete
genetic
  recombination
haploid
histone

homologous
  chromosome
interphase
karyotype
leukemia
life cycle
lymphoma
malignant tumor
meiosis
metaphase
metastasis
mitosis
mitotic phase
mitotic spindle
nucleosome

nondisjunction
polyploid
prophase
radiation therapy
sarcoma
sex chromosome
sexual
  reproduction
sister chromatid
somatic cell
telophase
tetrad
trisomy 21
tumor
zygote

Crossword Puzzle

Use the Key Terms list from this chapter to fill in the crossword puzzle.

ACROSS

1. final mitotic stage when the nuclear envelope reforms
3. type of cancer therapy exposing parts of the body to high energy
5. cancer that arises in tissues that support the body
6. a bead-like structure consisting of DNA wrapped around histone molecules
9. cancer of the tissue that forms white blood cells
11. a condition in which a person has three number 21 chromosomes
12. part of cell cycle when cell is dividing
16. time when a cell metabolizes and performs its various functions
17. mitotic stage when the chromosomes are lined up in the cell's middle
20. the division of a diploid cell into two diploid cells
22. type of chromosome that forms a pair
24. stages leading from the adults of one generation to the adults of the next
26. a display of the chromosomes of a cell
27. football-shaped structure of microtubules important in mitosis

28. long strands of DNA attached to certain proteins
30. trisomy 21
31. type of reproduction requiring fertilization
32. a thread-like, gene-carrying structure formed from chromatin
33. small protein associated with DNA
35. cells that contain two homologous sets of chromosomes
38. cell reproduction
39. egg or sperm
40. sequence of events from cell formation to cell division
43. type of tumor that stays at its original site of origin
44. the orientation of homologous pairs of chromosomes
45. a chromosome that does not determine sex
46. cloud of cytoplasmic material with centrioles in animal cells
47. type of reproduction not involving fertilization
48. indentation at the equator of a cell where it will divide
49. production of gene combinations unlike the parent chromosomes

DOWN

2. mitotic stage when the nuclear membrane first breaks up

4. the fertilized egg

7. type of chromosome that determines the sex of a child

8. the spread of cancer cells beyond their original site

9. cancer of blood-forming tissues

10. one of two identical parts of a replicated chromosome

11. an abnormally growing mass of body cells

13. process of fusion of haploid sperm and haploid egg cell

14. having more than two sets of homologous chromosomes in each somatic cell

15. the division of a diploid nucleus into four haploid daughter nuclei

18. type of cell with 46 chromosomes in humans

19. exchange of genetic material between homologous chromosomes

21. the result when paired chromosomes fail to separate

23. type of tumor that migrates away from its site of origin

25. the use of drugs to disrupt the division of cancer cells

29. division of the cytoplasm

33. cells that contain only one chromosome of each homologous pair

34. disease of cells that divide excessively and exhibit bizarre behavior

36. membranous disk containing cell wall material in plants

37. the region where two chromatids are joined

40. site where crossing over has occurred

41. cancer originating in the external or internal coverings of the body

42. mitotic stage when sister chromatids separate

Patterns of Inheritance

Studying Advice

a. Have you ever wondered how your sex was determined, why people have different skin colors, and how diseases such as hemophilia and sickle-cell disease are inherited? This chapter addresses some of the most interesting questions and relevant information in all of your biological studies!

b. This chapter includes many examples of different types of inheritance. To help you organize these definitions, Table 9-2 provides space for you to define and describe key terms and phrases related to inheritance.

c. If you have not recently studied chapter 8, you will need to review the process of meiosis to best understand the chapter 9 section titled "The Chromosomal Basis of Inheritance."

Student Media

Activities

Monohybrid Cross
Dihybrid Cross
Gregor's Garden
Incomplete Dominance
Linked Genes and Crossing Over
Sex-Linked Genes

Case Study in the Process of Science

What Can Fruit Flies Reveal About Inheritance?

MP3 Tutor

Chromosomal Basis of Inheritance

Videos

Ultrasound of Human Fetus 1
Ultrasound of Human Fetus 2
Discovery Channel Video Clip: Colored Cotton
Discovery Channel Video Clip: Novelty Gene

Organizing Tables

The table below shows the results of a cross between plants heterozygous for purple flower color. Write the genotypes of the plants in each square as either (a) homozygous dominant, (b) heterozygous, or (c) homozygous recessive. Next, write the phenotypes of the flowers for each square as either (a) purple (the dominant trait) or (b) white (the recessive trait).

TABLE 9.1	
Genotype: *PP*	Genotype: *Pp*
Genotype in words _____	Genotype in words _____
Phenotype _____	Phenotype _____
Genotype: *pP*	Genotype: *pp*
Genotype in words _____	Genotype in words _____
Phenotype _____	Phenotype _____

Organizing the key terms related to inheritance: This table should help you organize the definitions of key aspects of inheritance.

TABLE 9.2	
Term	Definition
Recessive disorders	
Dominant disorders	
Incomplete dominance	
Multiple-allele inheritance	
Codominance	
Pleiotropy	
Polygenic inheritance	
Linked genes	
Sex-linked genes	

Content Quiz

Directions: Identify the *one* best answer for the multiple-choice questions. For true/false questions, determine if the statement is true or false. If false, change the underlined word(s) to make the statement true. Finally, add the correct word(s) to the fill-in-the-blank questions to make the statements true.

Biology and Society: Testing Before Birth

1. When amniocentesis is performed:
   A. a piece of placenta is removed.
   B. sound waves are used to view the fetus.
   C. the fetus is viewed directly through a thin tube inserted into the uterus.
   D. amniotic fluid is sampled.

2. True or False? <u>Amniocentesis or chorionic villus sampling</u> could be used to test for Down syndrome.

3. Amniocentesis and chorionic villus sampling require the collection of _____.

Heritable Variation and Patterns of Inheritance

IN AN ABBEY GARDEN

4. Crossing members of two different true-breeding varieties of organisms produces:
   A. F_1 hybrids.
   B. F_2 hybrids.
   C. P_1 hybrids.
   D. P_2 hybrids.
   E. F_1 true breeders.

5. True or False?   When Mendel wanted to <u>cross-fertilize</u> his pea plants, he covered the flower with a small bag.

6. Fertilization between members of the F_1 generation produces members of the _____ generation.

MENDEL'S LAW OF SEGREGATION

7. What are the phenotypes of two pea plants with the flower-color genotypes *Pp* and *PP*?
   A. Both are purple.
   B. Both are white.
   C. One is white and the other is purple.
   D. A Punnett square is needed to figure this out.

8. If we crossed two pea plants with genotypes *Pp* and *PP* as noted above, we would expect that the phenotypes of the offspring would be:
   A. all purple.
   B. half purple and half white.
   C. three-quarters purple and one-quarter white.
   D. three-quarters white and one-quarter purple.
   E. all white.

9. Mendel's principle of segregation states that:
   A. pairs of alleles stay together during gamete formation and the alleles separate at fertilization.
   B. pairs of alleles segregate during gamete formation and the alleles separate at fertilization.
   C. pairs of alleles stay together during gamete formation and the alleles pair again at fertilization.
   D. pairs of alleles segregate during gamete formation and the alleles pair again at fertilization.

10. True or False?   An individual that has two different alleles for a gene is said to be <u>homozygous</u>.

11. The physical location of a gene on a chromosome is that gene's _____.

MENDEL'S LAW OF INDEPENDENT ASSORTMENT

12. If Mendel's principle of independent assortment *did not apply* to the pea shape and color experiment, and the two traits in a dihybrid cross of heterozygous individuals *were* inherited together, what would be the expected phenotypic ratio of the F_2 generation?
    A. 9:3:3:1
    B. 1:6:1
    C. 1:2:1
    D. 3:1
    E. 1:1

13. True or False?   Mendel performed dihybrid crosses involving all seven of his pea characteristics and found 9:3:3:1 ratios in <u>half</u> of them.

14. Mendel's pea shape and color experiment showed that this dihybrid cross was the equivalent of two _____ crosses occurring simultaneously.

USING A TESTCROSS TO DETERMINE AN UNKNOWN GENOTYPE

15. A testcross is a mating between:
    A. an organism of unknown genotype and a homozygous dominant individual.
    B. an organism of unknown genotype and a heterozygous individual.
    C. an organism of unknown genotype and a homozygous recessive individual.
    D. two heterozygous individuals.
    E. a homozygous dominant and a homozygous recessive individual.

16. True or False?   Mendel used <u>testcrosses</u> to determine whether he had true-breeding varieties of plants.

17. Testcrosses are used to determine the _____ of an organism.

THE RULES OF PROBABILITY

18. Using a standard deck of 52 playing cards, the chance of drawing a red card is 1/2 and the chance of drawing a queen is 1/13. To determine the probability of drawing a red queen, we should use the rule of:
    A. addition.
    B. subtraction.
    C. multiplication.
    D. division.

19. True or False?   If we know the <u>phenotypes</u> of the parents, we can predict the probability for any genotype among the offspring.

FAMILY PEDIGREES

20. Consider a trait in which T represents the dominant allele and t represents the recessive allele. Which of the following genotypes will exhibit the recessive phenotype?
    A. TT
    B. Tt
    C. tT
    D. tt
    E. more than one of the above
    F. none of the above

21. Examine the patterns of inheritance of deafness in Figure 9.13 of your text. How many of the seven children of Jonathan and Elizabeth are carriers for the deafness trait?
    A. 0
    B. 2
    C. 5
    D. 7

22. True or False?   Dominant phenotypes <u>are</u> more common than recessive phenotypes.

23. True or False?   A family <u>pedigree</u> is used to assemble information about the occurrence of heritable characteristics in parents and their offspring across several generations.

24. People who are heterozygous with just one copy of an allele for a recessive disorder do not show symptoms of the disorder. They are considered to be _____ of the disorder.

HUMAN DISORDERS CONTROLLED BY A SINGLE GENE

25. Dominant lethal alleles:
    A. are more common than lethal recessive alleles.
    B. are commonly carried by heterozygotes without affecting them.
    C. are never inherited from a heterozygote.
    D. cause most human genetic disorders.
    E. cause Huntington's disease.

26. True or False?   Inbreeding is <u>less</u> likely to produce offspring that are homozygous for a harmful recessive trait.

27. Most people born with recessive disorders are born to parents who are both _____ or carriers for the recessive allele.

28. The illness called _____ is a degeneration of the nervous system that usually does not begin until middle age.

29. The most common lethal genetic disease in the United States is _____.

Variations on Mendel's Laws
INCOMPLETE DOMINANCE IN PLANTS AND PEOPLE

30. If a mating occurs between two parent plants that are both heterozygous for a trait with incomplete dominance, the expected ratio of phenotypes will be:
    A. 1:3.
    B. 1:2:1.
    C. 1:1.
    D. 9:3:3:1.
    E. 2:1.

31. True or False?   Mendel's work involved plants that <u>showed</u> incomplete dominance.

32. Humans with the disease _____ are homozygous or heterozygous for an allele that causes dangerously high cholesterol levels in the blood.

ABO BLOOD TYPE: AN EXAMPLE OF MULTIPLE ALLELES AND CODOMINANCE

33. What is the expected *phenotypic* ratio of children from parents with the following blood genotypes? $I^A I^A$ and $I^B I^i$?
    A. 1/2 AB and 1/2 A
    B. 1/2 AB and 1/2 B
    C. 1/4 A, 1/2 AB, and 1/4 B
    D. 1/2 AB and 1/2 O
    E. 1/4 B, 1/2 A, and 1/4 O

34. Examine the blood test results in Figure 9.19 of your text. Why aren't any of the blood cells clumped in the test of AB blood?
    A. There must have been a mistake.
    B. AB blood has antibodies to type A and type B carbohydrates.
    C. AB blood does not have antibodies to type A or type B carbohydrates.
    D. AB blood has antibodies to type A carbohydrates.
    E. AB blood has antibodies to type B carbohydrates.

35. True or False?   When a trait exhibits codominance, both <u>recessive</u> phenotypes are expressed in heterozygotes with two dominant alleles.

36. People with type _____ blood show codominance because both alleles are expressed.

PLEIOTROPY AND SICKLE-CELL DISEASE

37. In tropical Africa, resistance to malaria occurs in people who are:
    A. homozygous for the sickle-cell allele.
    B. heterozygous for the sickle-cell allele.
    C. homozygous for the non-sickle-cell allele.
    D. born with an extra chromosome containing the non-sickle-cell allele.

38. True or False? About one in every ten African Americans are <u>heterozygotes</u> for the sickle-cell trait.

39. Sickle-cell disease is an example of _____, when a single gene impacts more than one characteristic.

POLYGENIC INHERITANCE

40. In the example of polygenic inheritance described in the text, regarding human skin color, which of the following genotypes would produce the darkest skin?
    A. AAbbcc
    B. AaBbCc
    C. aabbcc
    D. AAbBcc
    E. AaBbCC

41. True or False? In polygenic inheritance of skin-color, only the <u>dominant</u> alleles contribute to darker skin color.

42. The opposite of pleiotropy, _____ involves two or more genes affecting a single characteristic.

THE ROLE OF ENVIRONMENT

43. If we examine a real human population for the skin-color phenotype, we would see more shades than just seven because:
    A. of the effects of environmental factors.
    B. some genes fade over time, producing light effects.
    C. there appear to be more genes involved in skin tone than have been described.
    D. skin color has no genetic component.

44. True or False? Height clearly <u>has</u> a large environmental component.

45. Although organisms result from a combination of genetic and environmental factors, only _____ factors are passed on to the next generation.

The Chromosomal Basis of Inheritance

46. Linked genes tend to be inherited together because:
    A. they affect the same characteristic.
    B. they determine different aspects of the same trait.
    C. they have similar structures.
    D. they are on the same chromosome.
    E. they have the same alleles.

47. Which one of the following processes can result in the separate inheritance of linked genes?
    A. polygenic inheritance
    B. crossing over
    C. incomplete dominance
    D. pleiotropy
    E. independent assortment

48. The probability of crossover between two linked genes is greatest when the genes are:
    A. located on separate chromosomes.
    B. close together on the same chromosome.
    C. farthest apart on the same chromosome.
    D. pleiotropic.
    E. dominant.

49. True or False?   Linked genes do not follow the typical patterns of inheritance.

50. True or False?   Researchers used recombination data to assign genes to relative positions on chromosomes to create linkage maps.

51. The _____ states that genes are located at specific positions on chromosomes and that chromosomal behavior during meiosis and fertilization accounts for inheritance patterns.

52. Early studies of the relationship between chromosome behavior and inheritance relied upon _____, often seen flying around overripe fruit.

Sex Chromosomes and Sex-Linked Genes

SEX DETERMINATION IN HUMANS AND FRUIT FLIES, SEX-LINKED GENES

53. Which one of the following statements about sex chromosomes is *false*?

    A. Humans have 44 autosomes and 2 sex chromosomes.

    B. In both humans and fruit flies, a male inherits an X and a Y sex chromosome.

    C. A gene called SRY, found on the Y chromosome, triggers testis development.

    D. Each human gamete contains both sex chromosomes.

    E. Most sex-linked genes unrelated to sex determination are found on the X chromosome.

54. True or False?   Any gene located on a sex chromosome is a <u>sex-linked gene</u>.

55. The _____ chromosomes of a human female are XX.

SEX-LINKED DISORDERS IN HUMANS

56. For a sex-linked recessive allele to be expressed a man would have to inherit:

    A. only one recessive allele, but a woman would have to inherit two.

    B. two recessive alleles, but a woman would have to inherit only one allele.

    C. two recessive alleles, just like a woman.

    D. only one recessive allele, just like a woman.

57. True or False?   Sex-linked human diseases <u>do not occur</u> in females.

58. True or False?   Red-green color blindness is a common sex-linked disorder involving <u>one</u> X-linked gene(s).

59. Factors involved in blood clotting are missing in people suffering from _____.

60. A sex-linked recessive disorder called _____ is a condition characterized by a progressive weakening and loss of muscle tissue.

Evolution Connection: The Telltale Y Chromosome

61. Which one of the following statements about the human Y chromosome is *false*? The human Y chromosome:
    A. is only about 1/3 the size of the X chromosome.
    B. carries only 1/100 as many genes as the X chromosome.
    C. carries most of the genes that code for female fertility.
    D. engages in limited crossing over with the X chromosome during meiosis.
    E. may have evolved from an X chromosome about 300 million years ago.

62. True or False?   Most of the DNA of the Y chromosome passes <u>intact</u> from father to son.

63. Studies of the _____ chromosome have been used to trace the evolution of humans.

Word Roots

co = together (codominance: phenotype in which both dominant alleles in a heterozygous individual are expressed)

di = two (dihybrid: a type of cross that mates varieties differing in two characteristics)

geno = offspring (genotype: an organism's genetic makeup)

hemo = blood; **philia** = love (hemophilia: a human genetic disease caused by excessive bleeding following an injury)

hetero = different (heterozygous: when an organism has different alleles for a gene)

homo = alike (homozygous: when an organism has the same alleles for a gene)

hyper = excessive (hypercholesterolemia: a condition of incomplete dominance resulting in elevated blood cholesterol levels)

mono = one (monohybrid: a type of cross between organisms that differ in only one trait)

pheno = appear (phenotype: an organism's physical traits)

pleio = more; **trop** = change (pleiotropy: when a single gene impacts more than one characteristic)

poly = many; **gen** = produce (polygenic: type of inheritance in which two or more genes affect a single trait)

Key Terms

ABO blood groups
achondroplasia
alleles
carrier
chromosome theory of inheritance
codominance
cross
cross-fertilization
dihybrid cross
dominant allele
Duchenne muscular dystrophy
F_1 generation
F_2 generation
genetics
genotype
hemophilia
heterozygous
homozygous
Huntington's disease
hybrid
hypercholesterolemia
inbreeding
incomplete dominance
law of segregation
linkage map
linked genes
loci
Mendel's law of independent assortment
monohybrid cross
pedigree
P generation
phenotype
pleiotropy
polygenic inheritance
Punnett square
recessive allele
recombination frequency
red-green color blindness
rule of multiplication
self-fertilize
sex-linked gene
sickle-cell disease
testcross
true-breeding
wild-type traits

Crossword Puzzle

Use the Key Terms list from this chapter to fill in the crossword puzzle.

ACROSS

5. the way to determine the probability that two independent events will both occur

6. a sex-linked recessive trait that causes excessive bleeding

7. when an organism has the same alleles for a gene

8. mating of an organism of unknown genotype with a homozygous recessive organism

9. type of inheritance in which two or more genes affect a single trait

11. any gene located on a sex chromosome

12. hybridization of F_1 organisms

13. the type of traits most often seen in nature

14. when an organism has different alleles for a gene

17. a form of muscular dystrophy characterized by a progressive weakening and loss of muscle tissue

18. the parental organisms

20. a type of color blindness commonly sex-linked in humans

21. a type of cross mating varieties differing in two characteristics

22. in a heterozygote, the type of allele that has no noticeable effect on the phenotype

23. the science of heredity

24. when a single gene impacts more than one characteristic

25. a device for predicting the results of a genetic cross

26. hybrid offspring

28. phenotype in which both dominant alleles in a heterozygous individual are expressed

29. a type of map based on the frequencies of recombinations during crossover

31. a type of cross between organisms that differ in only one trait

32. the three letters representing the three types of human blood alleles

34. the specific locations of genes on chromosomes

35. a type of degenerative disease of the nervous system caused by a dominant allele

36. a hybridization

39. a condition of incomplete dominance resulting in elevated blood cholesterol levels

40. results of a cross of close relatives

41. alternate forms of genes

42. family tree describing the occurrence of heritable characters in parents and offspring

43. type of genes that are located close together on the same chromosome

DOWN

1. production of offspring with inherited trait(s) identical to the parents

2. the percentage of recombinant offspring in a testcross

3. a type of disease characterized by malformed blood cells and other symptoms

4. the genetic makeup of an organism

10. fusion of sperm and egg from different organisms

15. the principle that sperm or eggs carry only one allele for each inherited characteristic

16. an organism's physical traits

19. the law that states that each pair of alleles sorts independently of the other pairs of alleles during gamete production

27. a form of dwarfism caused by a dominant allele

30. type of dominance in which the heterozygote phenotype is intermediate to the homozygous phenotypes

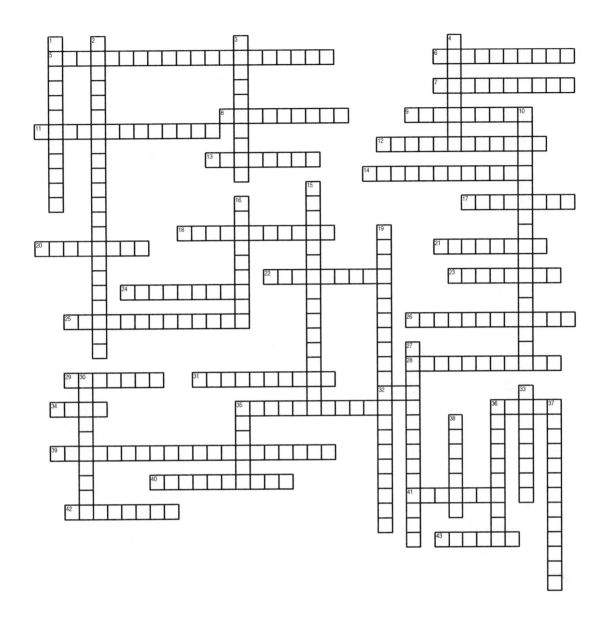

33. in a heterozygote, the type of allele that determines the phenotype

35. the offspring of two different true-breeding varieties

36. the type of theory of inheritance that states that genes are located on specific positions on chromosomes

37. fusion of sperm and egg from the same organism

38. an organism that has one allele for a recessive disorder but shows no symptoms

CHAPTER 10

The Structure and Function of DNA

Studying Advice

a. This chapter introduces the basics of molecular genetics, one of the most exciting and explosive fields of modern biology. A mastery of this chapter's content is necessary to understand the discussions in chapters 11 and 12.

b. Before reading this chapter, review the structure of nucleic acids in chapter 3.

c. Begin studying this chapter by examining Figure 10.9. This figure represents the overall flow of genetic information in a cell and is fundamental to the rest of the chapter.

Student Media

Activities

The Hershey-Chase Experiment
DNA and RNA Structure
DNA Double Helix
DNA Replication
Overview of Protein Synthesis
Transcription and RNA Processing
Translation
Simplified Reproductive Cycle of a Virus
Phage Lytic Cycle
Phage Lysogenic and Lytic Cycles
HIV Reproductive Cycle

Case Studies in the Process of Science

What Is the Correct Model for DNA Replication?
How Is a Metabolic Pathway Analyzed?

How Do You Diagnose a Genetic Disorder?
What Causes Infections in AIDS Patients?
Why Do AIDS Rates Differ Across the U.S.?

eTutor

Protein Synthesis

MP3 Tutor

Central Dogma

Video

Discovery Channel Video Clip: Emerging Diseases

Organizing Tables

Compare the structure of DNA and mRNA in the table that follows.

TABLE 10.1			
	Type of Sugar	Types of Bases	Overall Shape of the Molecule
DNA			
mRNA			

Compare the structures and functions of the different types of RNA molecules in the table below.

TABLE 10.2			
	Structure	Function	Location in the Cell
mRNA			
tRNA			
rRNA			

Content Quiz

Directions: Identify the *one* best answer for the multiple-choice questions. For true/false questions, determine if the statement is true or false. If false, change the underlined word(s) to make the statement true. Finally, add the correct word(s) to the fill-in-the-blank questions to make the statements true.

Biology and Society: Sabotaging HIV

1. Which one of the following statements is *false?*
   A. AIDS is one of the most significant health challenges facing the world today.
   B. The letters in AIDS stand for acquired immune deficiency syndrome.
   C. AIDS is spread through the exchange of body fluids.
   D. There is no cure for AIDS.
   E. HIV uses special enzymes to convert its DNA into RNA.

2. A molecule of AZT has a shape very similar to:
   A. ATP, an important molecule in energy transport inside cells.
   B. thymine, one of the four nucleotides that comprise DNA.
   C. water.
   D. ribosomal RNA.
   E. glucose, a common type of sugar found in cells.

3. True or False? AZT prevents the synthesis of HIV <u>RNA</u>.

DNA: Structure and Replication

4. In the 1950s, scientists understood the functions of DNA to be all of the following *except*:
   A. the capacity to *store* genetic information.
   B. the capacity to *create* new genetic information.
   C. the capacity to *copy* genetic information.
   D. the capacity to *pass along* genetic information from generation to generation.

5. True or False? Mendel worked on inheritance patterns <u>without knowing</u> about DNA's role in heredity.

6. By the 1950s, a race was on to understand the _____ of DNA.

DNA AND RNA STRUCTURE

7. The backbone of DNA and RNA polynucleotides consists of a repeating pattern of:
   A. sugar, base, sugar, base.

B. phosphate, base, phosphate, base.

C. sugar, phosphate, sugar, phosphate.

D. sugar, base, phosphate, sugar, base, phosphate.

8. True or False? The "D" in DNA comes from deoxyribose because compared to the sugar in RNA, the sugar in DNA <u>has an extra</u> oxygen atom.

9. The sugars of DNA and RNA are different. DNA has the sugar _____, and RNA has the sugar _____.

10. RNA uses the nitrogenous base _____, which DNA does not use.

11. The two types of bases in DNA are the single-ring bases, _____ and _____, and the double-ring structures, _____ and _____.

12. In DNA and RNA, the polymers are _____ and the monomers are _____.

WATSON AND CRICK'S DISCOVERY OF THE DOUBLE HELIX

13. The genetic information in a chromosome is encoded:

A. in the nucleotide sequence of the molecule.

B. by the base pairing of the DNA molecule.

C. by the interaction between DNA and the associated histone proteins.

D. in the types of sugars used in the molecule.

E. in the types of chemical bonds formed between the bases forming DNA.

14. The shape of a DNA molecule is most like the shape of:

A. railroad tracks.

B. a spiral staircase.

C. the strings on a tennis racquet.

D. the letter X.

15. If adenine paired with guanine and cytosine paired with thymine in DNA, then:

A. DNA would have irregular widths along its length.

B. the DNA molecule would be much longer.

C. the DNA molecule would be much shorter.

D. the sequential information would be lost.

E. the DNA molecule would be circular.

16. True or False? Watson and Crick discovered that the backbone of DNA was located on the <u>inside</u> of the molecule.

17. If one side of a DNA molecule has the bases CGAT, the opposite side would have the bases _____.

18. When a DNA molecule is copied, how much of the original DNA is included in the new copy?

    A. none

    B. 25%

    C. 50%

    D. 75%

    E. 100%

19. DNA replication is most like:

    A. picking students to make two baseball teams.

    B. two people getting divorced followed by each person remarrying.

    C. splitting a large plant into two and planting each half.

    D. mixing vinegar and oil to make salad dressing.

20. True or False?   The DNA molecule of a eukaryotic chromosome has <u>a single replication origin</u>.

21. Covalent bonds between the nucleotides of a new DNA strand are made by the enzyme _____.

22. DNA repair is accomplished by the enzyme _____ and some of the proteins associated with DNA replication.

The Flow of Genetic Information from DNA to RNA to Protein
HOW AN ORGANISM'S GENOTYPE PRODUCES ITS PHENOTYPE

23. The Beadle and Tatum hypothesis about the function of genes is best stated as one gene– _____:

    A. one protein.

    B. one enzyme.

    C. one DNA.

    D. one polypeptide.

    E. one monomer.

24. Which one of the following best represents the flow of genetic information in a cell?

    A. RNA → transcription → DNA → translation → PROTEIN

    B. DNA → transcription → RNA → translation → PROTEIN

    C. DNA → transcription → PROTEIN → translation → RNA

    D. DNA → translation → RNA → transcription → PROTEIN

25. True or False?   The molecular basis of the phenotype lies in an organism's <u>DNA</u>.

26. An organism's _____ is its genetic makeup. An organism's specific traits are its _____.

27. Cells produce RNA by the process of _____ and proteins by the process of _____.

FROM NUCLEOTIDES TO AMINO ACIDS: AN OVERVIEW

28. If a gene consisted of 60 bases, the protein made from the gene would be about:
    A. 180 amino acids long.
    B. 60 amino acids long.
    C. 20 amino acids long.
    D. 10 amino acids long.

29. True or False?   A DNA molecule may contain <u>thousands</u> of genes.

30. True or False?   A gene may consist of <u>thousands</u> of nucleotides.

31. A codon consists of _____ base(s) in a DNA or RNA molecule.

32. The sequence of nucleotides of the RNA molecule dictates the sequence of _____ of the polypeptide.

THE GENETIC CODE

33. Of the 64 possible codons:
    A. 61 code for an amino acid and 3 are stop codons.
    B. all of them code for at least one amino acid.
    C. 20 code for 2 amino acids and the rest code for a single amino acid.
    D. 2 are start codons, 3 are stop codons, and 59 code for a single amino acid.
    E. 1 is a start codon, 1 is a stop codon, and the remaining 62 code for a single amino acid.

34. Examine the sets of codons in Figure 10.11 that code for the same amino acid. How do the codons that all code for the same amino acid compare?
    A. The first letter in the codon is most likely to be different.
    B. The second letter in the codon is most likely to be different.
    C. The third letter in the codon is most likely to be different.
    D. The second and third letters in the codon are always the same.
    E. The first and third letters in the codon are always the same.

35. True or False?   There are <u>gaps</u> between the codons of RNA.

36. True or False?   Different codons may code <u>for the same amino acid</u>.

37. The set of rules relating nucleotide sequence to amino acid sequence is the _____.

TRANSCRIPTION: FROM DNA TO RNA

38. Which one of the following statements about transcription is *false*?
    A. In RNA, U, rather than T, pairs with A.
    B. The RNA molecule is built one nucleotide at a time.
    C. Both DNA strands serve as the template for one RNA.
    D. Transcription begins when RNA polymerase attaches to the promoter.
    E. As the RNA molecule is produced, it peels away from its DNA template.

39. True or False?   The terminator sequence indicates the <u>end</u> of a gene.

40. The _____ dictates which of the two DNA strands is to be transcribed.

41. The RNA nucleotides are linked by the enzyme _____.

THE PROCESSING OF EUKARYOTIC RNA

42. In eukaryotic cells, RNA transcribed from DNA undergoes processing before leaving the nucleus. In this processing:
    A. a cap and tail are added, introns are edited out, and exons are joined together to make mRNA.
    B. a cap and tail are removed, introns are edited out, and exons are joined together to make mRNA.
    C. a cap and tail are added, exons are edited out, and introns are joined together to make mRNA.
    D. a cap is removed, a tail is added, exons are edited out, and introns are joined together to make mRNA.

43. True or False?   The cap and tail on an mRNA molecule <u>protects mRNA from cellular enzymes</u> and helps ribosomes recognize RNA as mRNA.

44. RNA splicing involves the removal of _____ and the joining of _____ to produce mRNA.

TRANSLATION: THE PLAYERS

45. The actual translation of codons into amino acids is the job of:
    A. mRNA.
    B. tRNA.
    C. rRNA.
    D. ribosomes.
    E. RNA polymerase.

46. Which one of the following statements about tRNA is *false*?
    A. An anticodon recognizes a particular mRNA codon by using base-pairing rules.
    B. There is a slightly different version of tRNA for each amino acid.
    C. Some parts of the tRNA molecule twists, folds around, and base-pairs with itself.
    D. Each tRNA molecule must pick up the appropriate amino acid.
    E. A tRNA molecule is made of a double strand of RNA about 800 nucleotides long.

47. A ribosome consists of:
    A. one subunit made up of proteins and DNA.
    B. two subunits, each made up of tRNA and rRNA.
    C. two subunits, each made up of proteins and mRNA.
    D. two subunits, each made up of proteins and rRNA.
    E. three subunits, each made up of proteins, mRNA, and tRNA.

48. True or False?   The part of a tRNA molecule that binds to a codon is a parallel codon.

49. A fully assembled ribosome has a binding site for _____ on its small subunit and a binding site for _____ on its large subunit.

TRANSLATION: THE PROCESS, REVIEW: DNA → RNA → PROTEIN

50. During the initiation stage of translation:
    A. a ribosome assembles with the mRNA and the initiator tRNA bearing the first amino acid.
    B. additional amino acids are brought in, one at a time, as a polynucleotide forms.
    C. the ribosomal subunits form by the combination of rRNA and proteins.
    D. the mRNA molecule is edited further, beginning with the removal of the cap and tail.

51. Which one of the following is the correct molecular sequence in the combined processes of transcription and translation.
    A. DNA → mRNA → polypeptide → protein
    B. mRNA → DNA → polypeptide → protein
    C. DNA → polypeptide → mRNA → protein
    D. DNA → mRNA → protein → polypeptide
    E. mRNA → polypeptide → DNA → protein

52. True or False?   An incoming tRNA molecule first binds with the mRNA codon <u>at the A site</u>.

53. After a new amino acid is brought in, the polypeptide leaves the tRNA in the P site and attaches to the _____ on the tRNA in the A site.

54. Once a new amino acid is added to a polypeptide, the tRNA and mRNA move together from the _____ site to the _____ site. This process is known as _____.

55. Elongation continues until a(n) _____ reaches the ribosome's A site.

MUTATIONS

56. Compare the two sentences below.

    The cat hit the red toy pig.

    The caz thi tth ere dto ypi.

    The second sentence has been changed in a way that is most like a mutation caused by:

    A. a base substitution.

    B. a single base deletion.

    C. a single base addition.

    D. a multiple base deletion.

    E. a multiple base addition.

57. In the mutation above, the second sentence no longer reads properly because:

    A. of a shift in the reading frame.

    B. most of the letters are different.

    C. several key letters were lost.

    D. of word substitutions.

    E. of word deletions.

58. True or False?   Alleles frequently differ by only <u>a single</u> base pair.

59. True or False?   A base substitution <u>shifts</u> the reading frame.

60. True or False?   Mutations are usually <u>beneficial</u>.

61. Any change in the nucleotide sequence of DNA is a(n) _____.

62. Chemicals and X-rays that cause mutations are called _____.

Viruses: Genes in Packages

BACTERIOPHAGES

63. In the lytic cycle of a bacteriophage:
    A. the DNA inserts by genetic recombination into the bacterial chromosome.
    B. the phage genes remain inactive.
    C. the viral DNA can be passed along through many generations of infected bacteria.
    D. the DNA immediately turns the cell into a virus-producing factory.
    E. the phage DNA is referred to as a prophage.

64. Which one of the following is most like the way that phage DNA is replicated during a lysogenic cycle?
    A. having a friend photocopy your two-page essay while he is photocopying a 50-page article
    B. having your friend wash a shirt for you while she washes her clothes
    C. asking a friend to help you while you change the tires on your truck
    D. inviting friends over to your house to share a good home-cooked meal
    E. working with a group of friends to design an experiment for a science class

65. True or False? The <u>lysogenic</u> cycle leads to the lysis of the host bacterial cell.

66. True or False? Once a prophage forms, it <u>cannot</u> leave its chromosome.

67. True or False? Prophage DNA is replicated <u>along with the host cell's DNA</u>.

68. Phage DNA that is incorporated into the bacterial chromosome is referred to as a(n) _____.

PLANT VIRUSES, ANIMAL VIRUSES

69. Which one of the following statements about plant viruses is *false*?
    A. Most plant viruses discovered to date have RNA instead of DNA as their genetic material.
    B. Most plant viruses consist of RNA surrounded by proteins.
    C. Plant viruses cannot easily penetrate a healthy plant epidermis.
    D. Plants infected with viruses can pass the virus on to their offspring.
    E. Unlike for animals, there are many simple cures for viral diseases of plants.

70. Which one of the following does *not* typically occur during the reproductive cycle of an enveloped RNA virus?
    A. A protein-coated RNA enters the host cell. Once inside the host cell, the protein coat around the virus is removed.
    B. A viral enzyme starts making complementary strands of the viral RNA.
    C. Some of the new RNA serves as mRNA for the synthesis of new viral proteins.
    D. The new viral proteins assemble around new viral RNA.
    E. The new viruses escape by killing the cells.

71. True or False?   Antibiotics <u>are</u> effective for the treatment of viral infections.

72. True or False?   We usually recover from colds by <u>replacing cells damaged by the virus</u>.

73. The herpes virus DNA may remain as _____ in the nuclei of certain nerve cells.

74. Most RNA viruses get their outer membranes from the _____ of the host cell. But DNA viruses, such as the herpes virus, get their outer membranes from the _____ of the host cell.

HIV, THE AIDS VIRUS

75. Which one of the following statements about HIV is *false*?
    A. The virus that is transmitted into cells contains RNA.
    B. Once in a host cell, the viral RNA synthesizes more RNA from the host's DNA.
    C. Double-stranded DNA produced by reverse transcription is inserted into the host cell's chromosomal DNA as a provirus.
    D. The provirus is transcribed and translated into viral proteins.
    E. New HIV leaves without killing the host cell.

76. The outer envelope of HIV is derived from the _____ of a previous host cell.

77. HIV is an example of a(n) _____, a virus that reproduces by means of a DNA molecule.

78. HIV uses the enzyme _____ to catalyze reverse transcription.

Evolution Connection: Emerging Viruses

79. RNA viruses mutate more quickly than our DNA because RNA viruses:

    A. come into contact with many mutagens.

    B. are made of more fragile amino acids.

    C. are smaller.

    D. lack proofreading steps after replication.

    E. have weaker covalent bonds between their bases.

80. True or False? Annual flu vaccines are necessary because last year's flu <u>virus may be mutated enough</u> that you have little immunity against it.

81. True or False? New viral diseases may result when an old virus is introduced to a new <u>host</u>.

82. Viruses that have appeared suddenly or may have only recently come to the attention of science are called _____ viruses.

Word Roots

muta = change; **gen** = producing (mutagen: a physical or chemical agent that causes mutations)

phage = eat (bacteriophages: viruses that attack bacteria)

poly = many (polynucleotide: a polymer of many nucleotides)

pro = before (prophage: phage DNA inserted into the bacterial chromosome before viral replication)

retro = backward (retrovirus: an RNA virus that reproduces by first transcribing its RNA into DNA then inserting the DNA molecule into a host DNA)

trans = across; **script** = write (transcription: the transfer of genetic information from DNA into an RNA molecule)

Key Terms

adenine (A)
AIDS
bacteriophages
cap
codon
cytosine (C)
DNA
DNA polymerase
double helix
emerging viruses
exons
genetic code

guanine (G)
HIV
introns
lysogenic cycle
lytic cycle
messenger RNA
molecular biology
mutagen
mutation
nucleotide
phages
polynucleotide

prophage
provirus
reading frame
retrovirus
reverse
  transcriptase
ribosomal RNA
  (rRNA)
RNA polymerase
RNA splicing
stop codon

sugar-phosphate
  backbone
tail
terminator
thymine (T)
transcription
transfer RNA
  (tRNA)
translation
uracil (U)
virus

Crossword Puzzle

Use the Key Terms list from this chapter to fill in the crossword puzzle.

ACROSS

1. the nitrogenous base that pairs with adenine
2. the transfer of genetic information from RNA into a protein
3. a bit of nucleic acid wrapped in a protein coat, these can cause serious disease
4. an RNA virus that reproduces by means of a DNA molecule
7. the selective editing out of introns
9. the transcription enzyme used to link RNA nucleotides
10. the transfer of genetic information from DNA into an RNA molecule
13. the study of heredity at the molecular level
17. the abbreviation for the type of polynucleotide transcribed from a gene
19. the nitrogenous base found only in RNA
20. a triplet that signals for translation to stop
21. a type of viral replication cycle in which viral DNA replication occurs without phage production
23. a physical or chemical agent that causes mutations
24. viral DNA that inserts into a host genome
25. the set of rules relating nucleotide sequence to amino acid sequence
28. the enzyme that makes the covalent bonds during DNA replication
30. a monomer of a nucleic acid
32. the abbreviation for acquired immune deficiency syndrome
33. viruses that attack bacteria
36. phage DNA inserted into the bacterial chromosome
37. the nitrogenous base that pairs with cytosine
38. an enzyme that catalyzes reverse transcription

DOWN

1. a nucleotide sequence that signals the end of a gene
4. the triplet grouping of an mRNA molecule
5. the abbreviation for the type of polynucleotide that helps to form ribosomes
6. a polymer of nucleotides
8. the abbreviation for the type of polynucleotide that serves as a molecular interpreter
11. extra nucleotides added to the end of an RNA transcript
12. a type of viral replication cycle that releases new phages by the death of the host cell
14. the short name for viruses that attack bacteria
15. noncoding stretch of nucleotides in RNA transcripts
16. a change in the nucleotide sequence of DNA
18. the general shape of a DNA molecule
20. the type of structural backbone of polynucleotides
22. the type of virus that has appeared suddenly or recently come to scientific attention
26. coding stretches of nucleotides in RNA transcripts
27. the nitrogenous base that pairs with guanine
29. the three-base word used during translation
31. the abbreviation for the double-stranded nucleic acid stored in the nucleus
32. the nitrogenous base that pairs with thymine
34. extra nucleotides added to the beginning of an RNA transcript
35. the abbreviation for the human immunodeficiency virus

CHAPTER 11

How Genes Are Controlled

Studying Advice

a. The content of this chapter provides great depth and understanding to big questions in biology. For example: (a) What causes cancer and *what can you do* to reduce your chances of getting cancer? (b) How do our cells specialize and arrange themselves as we develop? (c) How can we clone animals and why would we want to do this? and (d) What are stem cells and how can they help treat disease? Although the material is challenging, it is a chance to understand the cutting edge of research in these important areas of biology and medicine.

b. Use the two organizing tables below to "keep track" of the details in the chapter. The tables can serve as important reference points as you read further and review for exams.

Student Media

Activities

The *lac* Operon in *E. coli*

Gene Regulation in Eukaryotes

Review: Gene Regulation in Eukaryotes

Signal-Transduction Pathway

Causes of Cancer

Case Study in the Process of Science

How Do You Design a Gene Expression System?

MP3 Tutor

Control of Gene Expression

Videos

Biotechnology Lab
Discovery Channel Video Clip: Cloning
Discovery Channel Video Clip: Fighting Cancer

You Decide

Do Cell Phones Cause Brain Cancer?
Is Second-hand Smoke Dangerous?

Organizing Tables

Using the following table, describe the main functions of each component of the lactose (*lac*) operon and the regulatory genes and repressor proteins that affect the operon. Some of the cells of the table are already filled in to make this task a bit easier.

TABLE 11.1			
	Location	Function	What Controls Them
Promoter			
Operator			
Three enzyme genes	All three enzyme genes are located together next to the operator.		
Regulatory gene		Produces the repressor protein.	Nothing regulates it. It is always on!
Repressor protein	The repressor proteins are free-floating in the cytoplasm.		Lactose, when present, binds and prevents the repressor from attaching to the promoter.

Use the following table to compare the genes and process of gene regulation in prokaryotes and eukaryotes. Some of the cells of the table are already filled in to make this task a bit easier.

TABLE 11.2		
	Prokaryotes	Eukaryotes
Which group(s) use regulatory proteins that attach to DNA?		
Which group uses genes with their own promoter and control sequences?		
Which are used more often, activators or repressors?		
Are the operators (prokaryotes) or enhancers (eukaryotes) far away or close to the genes they help regulate?		
Which group(s) modify mRNA before it is translated?	Prokaryotes do not usually modify mRNA.	
How long do the mRNA molecules last in the cell?	Prokaryotic mRNA is degraded by enzymes after only a few minutes.	

Content Quiz

Directions: Identify the *one* best answer for the multiple-choice questions. For true/false questions, determine if the statement is true or false. If false, change the underlined word(s) to make the statement true. Finally, add the correct word(s) to the fill-in-the-blank questions to make the statements true.

Biology and Society: Cloning at the Edge of Extinction

1. If adult body cells typically contain the same genetic material, how do different types of cells develop?
   A. by turning off and on certain genes
   B. through the selective addition of some genes
   C. through the selective loss of some genes
   D. by joining cells together in unique combinations
   E. by mutations

2. True or False?   Cloned animals typically <u>have</u> been just as healthy as noncloned animals.

3. The nucleus of an adult body cell contains a complete _____ capable of directing the production of an entire organism.

How and Why Genes Are Regulated
PATTERNS OF GENE EXPRESSION IN DIFFERENTIATED CELLS

4. Genes determine the nucleotide sequences of:
   A. DNA.
   B. proteins.
   C. lipids.
   D. amino acids.
   E. mRNA.

5. The overall process by which genetic information flows from genes to proteins is called:
   A. replication.
   B. gene expression.
   C. gene suppression.
   D. transcription.
   E. translation.

6. True or False?   The genes for specialized proteins are expressed in <u>all</u> cells.

7. Individual cells must undergo _____; that is, they must become specialized in structure and function.

GENE REGULATION IN BACTERIA

8. In bacteria, gene expression is mainly controlled by:
   A. deleting certain genes from chromosomes.
   B. moving DNA into special capsules.
   C. limiting DNA replication.
   D. turning transcription on and off.
   E. making extra copies of chromosomes.

9. Examine Figure 11.5 in the text. A gene mutation would produce the greatest effects:
   A. when changes are made to the polypeptide in the cytoplasm.
   B. during translation.
   C. during processing of RNA in the nucleus.
   D. during transcription.

10. Which one of the following statements about the *lac* operon is *false*?
   A. Enzymes that help absorb and process lactose are produced by *E. coli* when lactose is absent.
   B. When RNA polymerase attaches to the promoter, it initiates transcription.
   C. The operator helps to determine whether RNA polymerase can attach to the promoter.
   D. The repressor protein binds to the operator and blocks the attachment of RNA polymerase to the promoter.
   E. When lactose is present, it interferes with the attachment of the *lac* repressor to the promoter by binding to the repressor and changing its shape.

11. The way that lactose works in the *lac* operon is most like:
   A. adding milk and sugar to coffee to improve its flavor.
   B. a boy distracting his mom while his brother takes some cookies.
   C. cooking a meal and serving it to guests.
   D. putting an ATM card into an ATM to get some money.
   E. advertising a restaurant to attract customers.

12. True or False?   RNA polymerase attaches to the <u>operator</u>.

13. A cluster of genes with related functions, along with the control sequences, is called a(n) _____.

GENE REGULATION IN EUKARYOTIC CELLS

14. The extra X chromosome in human females:
    A. is expressed at about the same level as the other X chromosome.
    B. is eliminated from the cell early in embryonic development.
    C. is highly compacted and inactivated.
    D. is less folded and more frequently expressed.

15. Eukaryotes usually:
    A. have operons.
    B. have a promoter and other control sequences for each gene.
    C. have regulatory proteins that bind to DNA.
    D. have more repressor genes than activators.

16. The process in Figure 11.8 is most like:
    A. baking a pie and a cake and cutting them up to serve to guests.
    B. clipping news and sports articles out of newspapers to make two scrap books.
    C. sorting out beads to make two different necklaces.
    D. editing 8 hours of film different ways to produce different movies.

17. In eukaryotes, the most important stage for regulating gene expression is the:
    A. unpacking of chromosomal DNA.
    B. breakdown of mRNA.
    C. removal of introns from RNA.
    D. initiation of transcription.
    E. transport of mRNA from the nucleus to the cytoplasm.

18. Which one of the following is *not* a mechanism used to regulate gene expression after eukaryotic mRNA is transported to the cytoplasm?
    A. The mRNA molecule is typically broken down within hours to weeks.
    B. Different mRNA molecules combine in the cytoplasm to form new mRNA molecules.
    C. Proteins in the cytoplasm regulate translation.
    D. Proteins are edited after translation.
    E. Some final protein products last only a few minutes or hours.

19. True or False?   DNA packing tends to <u>promote</u> gene expression.

20. True or False?   Eukaryotic cells have <u>more</u> elaborate mechanisms than bacteria for regulating the expression of their genes.

21. True or False?   Both prokaryotes and eukaryotes regulate transcription by using regulatory proteins that bind to <u>mRNA</u>.

22. Eukaryotic genes may be turned on when transcription factors bind to DNA sequences called _____.

23. Repressor proteins may bind to DNA sequences called _____, inhibiting the start of transcription.

CELL SIGNALING, DNA MICROARRAYS: VISUALIZING GENE EXPRESSION

24. When using a microarray, a researcher begins by collecting _____ from a particular type of cell.
    A. mRNA
    B. DNA
    C. proteins
    D. lipids
    E. ATP

25. True or False?   A DNA microarray consists of a glass slide containing thousands of different single-stranded <u>mRNA</u> fragments arranged in a grid.

26. True or False?   Cell-to-cell <u>signaling</u> is a key mechanism in development and in the coordination of cellular activities throughout an organism's life.

27. In cell-to-cell signaling, a signal molecule usually acts by binding to a receptor protein in the plasma membrane of the target cell and initiating a(n) _____ pathway.

28. Researchers can use microarrays to learn what _____ are active in different tissues.

Cloning Plants and Animals

THE GENETIC POTENTIAL OF CELLS, REPRODUCTIVE CLONING OF ANIMALS

29. In the process of nuclear transplantation, the nucleus from a donor cell is transplanted into:
    A. an egg in which the nucleus has been removed.
    B. a normal egg, allowing the two nuclei to fuse.
    C. the nucleus of another adult cell.
    D. another adult cell in which the nucleus has been removed.
    E. a sperm, which is used to fertilize an egg.

30. Reproductive cloning can be used to:
    A. produce herds of farm animals with desired traits.
    B. restock populations of endangered animals.

C. produce pigs for organ donation that lack a gene that produces a protein that can cause immune system rejection in humans.

D. all of the above.

E. none of the above.

31. True or False?   Hundreds or thousands of genetically identical <u>clones</u> can be produced from the cells of a single plant.

32. True or False?   The process of cloning shows that differentiation <u>does not</u> involve irreversible changes in the DNA.

33. True or False?   Dolly, the first mammal to be cloned, genetically resembled the <u>egg</u> donor.

34. Salamanders are capable of _____, the regrowth of lost body parts.

35. True or False?   Animal cloning was first performed in the <u>1990s</u>.

THERAPEUTIC CLONING AND STEM CELLS

36. Which one of the following statements is *false*?

A. When grown in laboratory culture, embryonic stem cells can divide indefinitely.

B. Adult stem cells are much more difficult than embryonic stem cells to grow in culture.

C. If the right conditions are used, scientists can induce changes in gene expression that cause differentiation of embryonic stem cells into a particular cell type.

D. In the future, embryos may be created using a cell nucleus from a patient so that ES cells can be harvested and induced to develop into replacement tissues or organs.

E. Embryonic stem cells are partway along the road to differentiation and usually give rise to only a few related types of specialized cells.

37. True or False?   <u>Embryonic</u> stem cells generate replacements for nondividing differentiated cells in adults.

38. The purpose of _____ cloning is to produce embryonic stem cells.

The Genetic Basis of Cancer

GENES THAT CAUSE CANCER

39. Which one of the following does *not* typically promote cancer?

A. bacteria transmitting tumor-suppressor genes

B. mutations in proto-oncogenes that code for growth factors

C. the inactivation of tumor-suppressor genes that inhibit cellular growth

D. a mutation that causes the *ras* protein to be hyperactive

40. Cancers usually take a long time to develop because:
    A. oncogenes are rare.
    B. only old cells can become cancerous.
    C. several specific mutations must occur.
    D. cancer cells usually grow very slowly.

41. Finding a single cure for all cancer is unlikely because:
    A. we've made so little progress in recent years.
    B. we just don't understand the genetics of the disease.
    C. the rapidly dividing cells cannot be killed.
    D. cancer is caused by many different factors.
    E. cancer cells migrate throughout the body.

42. True or False?   Some viruses carry <u>oncogenes</u>.

43. True or False?   For a proto-oncogene to become an oncogene, a cell's <u>RNA</u> must become mutated.

44. True or False?   Cancer <u>always</u> results from changes in DNA.

45. True or False?   Most cases of breast cancer are <u>caused</u> by inherited mutations.

46. Many proto-oncogenes code for _____, proteins that stimulate cell division.

CANCER RISK AND PREVENTION

47. The one substance known to cause more types of cancer is:
    A. alcohol.
    B. asbestos.
    C. X-rays.
    D. table salt.
    E. tobacco.

48. Which one of the following is *not* a cancer risk factor?
    A. a diet high in fat
    B. a diet high in protein
    C. the use of tobacco
    D. exposure to ultraviolet light
    E. alcohol consumption

49. True or False?   People can reduce their risks of developing colon cancer by consuming a diet high in <u>sugar</u>.

50. Cancer-causing compounds called _____ include ultraviolet light and tobacco smoke.

Evolution Connection: Homeotic Genes

51. Homeoboxes, found in homeotic genes:

    A. are types of proto-oncogenes that are a common cause of breast cancer.

    B. promote cancer by increasing the rate of cellular division.

    C. are characteristic of only mammals and birds.

    D. appear to be of recent evolutionary origin, coding for many animal traits that have only recently evolved.

    E. are very similar in many diverse organisms, suggesting a common evolutionary heritage.

52. True or False?   Homeoboxes containing homeotic genes have <u>different</u> developmental roles in mice and fruit flies.

53. Researchers studying homeotic genes found a common sequence of 180 _____.

Word Roots

homeo = alike (homeobox: a 180-nucleotide sequence within a homeotic gene)
onkos = tumor (oncogene: a gene that causes cancer)
proto = first (proto-oncogene: a normal gene with the potential to become an oncogene)

Key Terms

activators
adult stem cells
alternative RNA
  splicing
carcinogens
complementary
  DNA (cDNA)
cellular
  differentiation
clones
DNA microarray

embryonic stem
  cells (ES cells)
enhancers
gene expression
gene regulation
growth factors
homeoboxes
homeotic genes
nuclear
  transplantation
oncogene

operator
operon
promoter
proto-oncogene
regeneration
repressor
reproductive
  cloning
silencers

therapeutic
  cloning
transcription
  factors
tumor-suppressor
  genes
X chromosome
  inactivation

Crossword Puzzle

Use the Key Terms list from this chapter to fill in the crossword puzzle.

ACROSS

3. cancer-causing agent

5. using a somatic cell to make one or more genetically identical individuals

6. a glass slide containing thousands of kinds of single-stranded DNA fragments

7. the overall process by which genetic information flows from genes to proteins

8. a cancer-causing gene

10. a site where the transcription enzyme RNA polymerase attaches and initiates transcription

11. undifferentiated cells in an embryo that undergo unlimited division and produce several different types of cells

13. proteins secreted by certain cells that stimulate other cells to divide

15. DNA sequences where eukaryotic activators bind

16. a DNA molecule made *in vitro* using mRNA as a template and reverse transcriptase

17. the replacement of body parts

18. genetically identical organisms

22. proteins that switch on a gene or group of genes

24. a master control gene that regulates batteries of other genes

25. a cluster of genes with related functions along with a promoter and an operator

26. a 180-nucleotide sequence within a homeotic gene

DOWN

1. eukaryotic gene regulatory proteins

2. a molecule that can turn off transcription

4. a way for an organism to generate more than one polypeptide from a single gene

6. the process of cells becoming specialized in structure and function

9. a type of switch between the promoter and the enzyme genes

12. DNA sequence where eukaryotic repressor proteins bind

13. the turning on and off of genes

14. the process by which one of two X chromosomes is inactivated at random

19. a normal gene with the potential to become an oncogene

20. the type of gene that codes for proteins that normally help prevent uncontrolled cell growth

21. the type of cloning that scientists use to help patients with irreversibly damaged tissues

23. type of adult cell that generates replacements for nondividing differentiated cells

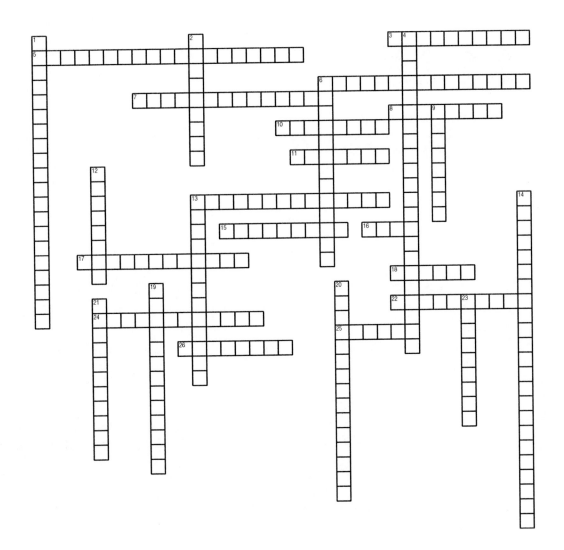

DNA Technology

Studying Advice

a. This is a long and challenging chapter, but one with plenty of rewards. The medical, legal, and social issues related to genetic engineering are now a common part of our national dialogue. This chapter addresses issues straight out of today's headlines.

b. If you have not read and understood the major content in chapters 10 and 11, review it before reading chapter 12. This chapter clearly builds on the information in these two prior chapters. The techniques described in chapter 12 require a good understanding of molecular genetics.

c. This is not a chapter to study for the first time the night before an exam. If your instructor will address parts or all of this chapter in lecture, be sure to read this chapter *before* attending the lectures. Take notes on the techniques as you read. Make your own diagrams detailing the steps of these procedures and create your own mini-glossary to refer to as you read further and listen in lecture. This is one of the hottest fields in all of science, affecting what food we eat, how we fight disease, and the way we view ourselves.

Student Media

Activities

Applications of DNA Technology

DNA Technology and Golden Rice

Cloning a Gene in Bacteria

Restriction Enzymes

DNA Fingerprinting

Gel Electrophoresis of DNA

Analyzing DNA Fragments Using Gel Electrophoresis

The Human Genome Project: Human Chromosome 17

Case Studies in the Process of Science

How Are Plasmids Introduced Into Bacterial Cells?

How Can Gel Electrophoresis Be Used to Analyze DNA?

Videos

Discovery Channel Video Clip: Transgenics
Discovery Channel Video Clip: DNA Forensics
Discovery Channel Video Clip: Colored Cotton

Organizing Tables

Indicate the starting materials and products of each of the procedures described in the text. To help you in this task, the "procedure" cells in the table are already completed.

TABLE 12.1

Process	Starting Materials	Procedure	Product
Cloning genes in a recombinant plasmid		1. Use the restriction enzyme to cleave the plasmid in only one place. 2. Use the same restriction enzyme to cleave the DNA into many pieces.	
Making pure genes using reverse transcriptase		1. Use reverse transcriptase to make a pure gene. 2. Insert the pure gene into a bacterium.	

Describe the function of each of the enzymes listed in the table below.

TABLE 12.2

Enzyme	Enzyme Function
DNA ligase	
DNA polymerase	
Restriction enzymes	
Reverse transcriptase	

Content Quiz

Directions: Identify the *one* best answer for the multiple-choice questions. For true/false questions, determine if the statement is true or false. If false, change the underlined word(s) to make the statement true. Finally, add the correct word(s) to the fill-in-the-blank questions to make the statements true.

Biology and Society: Crime Scene Investigations: Murders in a Small Town

1. The first legal use of DNA technology revealed that:
   A. one man was innocent and another was guilty.
   B. two men were guilty of rape.
   C. two men were innocent of rape.
   D. there are genetic differences between identical twins.
   E. DNA fingerprinting is not a reliable technique.

2. True or False?   DNA technology is the study and manipulation of <u>protein</u> material.

3. The scientific analysis of evidence for legal investigations is called
   _____.

Recombinant DNA Technology

4. Recombinant DNA technology combines:
   A. genes from different sources.
   B. the nucleus of one cell with the cytoplasm of another.
   C. all of the genetic material of two cells.
   D. proteins from one cell and DNA from another.
   E. proteins from two different cells.

5. True or False?   A host that carries <u>recombinant DNA</u> is called a transgenic or genetically modified organism.

6. True or False?   Biotechnology was first used <u>hundreds</u> of years ago to make bread and wine.

7. Research on *E. coli* led to the development of _____, a set of laboratory techniques for combining genes from different sources into a single DNA molecule.

8. The use of organisms to perform practical tasks defines _____.

FROM HUMULIN TO GENETICALLY MODIFIED FOODS

9. Which one of the following, if any, is *not* produced by recombinant DNA technology?

    A. humulin

    B. human growth hormone

    C. vaccines

    D. insect-resistant plants

    E. All of the above are produced by recombinant DNA technology.

10. True or False? Today, about half of all American corn crops are genetically modified.

11. Genetic engineering has produced rice that can help prevent _____ deficiency, a disease that often leads to vision impairment and increases susceptibility to disease.

12. Genetic engineering has now produced _____ that synthesize and secrete humulin.

13. Genetic engineering has produced potatoes that may help provide immunity against the disease _____.

14. For many viral diseases, prevention by _____ is the only medical way to fight the disease.

RECOMBINANT DNA TECHNIQUES

Sequencing: Number the following seven steps in the order that they occur in the process of cloning genes in recombinant plasmids.

_____ 15. The recombinant DNA plasmids are mixed with bacteria. The bacteria take up the recombinant plasmids.

_____ 16. The plasmid and human DNA are cut.

_____ 17. The transgenic bacteria, with the desired gene, is grown in large tanks, producing large quantities of the desired protein.

_____ 18. A biologist isolates two kinds of DNA: many copies of a bacterial plasmid to serve as a vector and human DNA containing a gene of interest.

_____ 19. The bacterial clone with the specific gene of interest is identified.

_____ 20. The cut DNA is mixed. The human DNA and plasmids join together.

_____ 21. Each bacterium, with its recombinant plasmids, is allowed to reproduce.

22. When biologists want to customize bacteria to produce a specific protein, the gene for that protein is typically inserted into:

    A. the chromosome of another bacterium.

    B. the coat of a phage.

    C. the DNA of a phage.

    D. a plasmid.

    E. the chromosome of the bacterium.

23. True or False?   The DNA of a plasmid is <u>part of</u> the bacterial chromosome.

24. True or False?   A nucleic acid probe <u>can be used</u> when only a part of the nucleotide sequence of a gene is already known.

25. True or False?   In a genomic library, each cell within a clone carries <u>a different</u> recombinant plasmid.

26. True or False?   A <u>nucleic acid probe</u> can be used to identify a bacterial clone carrying a particular gene of interest amongst the thousands of clones produced by shotgun cloning.

27. A(n) _____ is a short, single-stranded molecule of radioactively labeled DNA whose nucleotide sequence is complementary to part of the gene or other DNA of interest.

28. The cutting tools for making recombinant DNA are bacterial enzymes called _____.

29. The entire collection of cloned DNA fragments from a shotgun experiment, in which the starting material is bulk DNA from whole cells, is called a(n) _____.

30. When plasmids function as DNA carriers, moving genes from one cell to another, they are acting as _____.

31. The workhorses of modern biotechnology are _____.

32. A recombinant DNA molecule is produced when DNA pieces are connected into a continuous strand by _____, which forms covalent bonds between adjacent nucleotides.

DNA Fingerprinting and Forensic Science
MURDER, PATERNITY, AND ANCIENT DNA

33. Which one of the following, if any, is *not* a typical step in the DNA fingerprinting process?

    A. DNA samples are collected from different sources.

    B. When the genetic material is insufficient to analyze, DNA in the sample is amplified.

    C. Proteins are produced from the DNA.

D. The DNA samples are compared to each other.

E. All of the above steps are used in a typical DNA fingerprinting process.

34. True or False?   DNA from semen can be compared to DNA <u>from blood</u> to help solve crimes.

35. Since its introduction in 1986, _____ has become a standard part of law enforcement.

DNA FINGERPRINTING TECHNIQUES

36. Which one of the following techniques is used to sort macromolecules, primarily on the basis of their electric charge and length?
    A. gel electrophoresis
    B. RFLP analysis
    C. recombinant DNA technology
    D. gene cloning
    E. polymerase chain reaction

37. The _____ is a technique by which any segment of DNA can be cloned.

38. Repetitive DNA sequences from different individuals can be compared using _____ analysis.

39. The key to PCR is an unusual _____ that can withstand the heat needed to separate DNA strands.

40. Much of the DNA between genes is _____ DNA, which can be used in STR analysis.

Genomics and Proteomics
THE HUMAN GENOME PROJECT

41. The first targets of genomics research were:
    A. algae.
    B. human cells.
    C. eukaryotic disease-causing microbes.
    D. prokaryotic disease-causing microbes.
    E. viruses.

42. Which one of the following is *false*?
    A. The human genome carries between 2,000 and 3,000 genes.
    B. The human genome contains approximately 2.9 billion nucleotide pairs.
    C. Much of the DNA of humans consists of introns and repetitive DNA.
    D. The genomes of many multicellular organisms have been sequenced.
    E. About 97% of the entire human genome consists of DNA that does not code for proteins.

43. True or False?   The human genome consists of 24 different types of <u>genes</u>.

44. The science of studying whole genomes is called _____.

45. Much of the DNA between genes consists of nucleotide sequences present in many copies, called _____.

TRACKING THE ANTHRAX KILLER

46. When the genomes of the anthrax spores used in the 2001 bioterrorist attacks were compared, it was determined that the mailed spores:
    A. were not identical.
    B. came from different sources.
    C. were all a harmless veterinary vaccine strain.
    D. were all from the Ames strain.
    E. all came from one particular laboratory.

47. True or False?   Comparative genomics has revealed that humans and chimpanzees share <u>about 65%</u> of their DNA.

GENOME-MAPPING TECHNIQUES

48. The stage of the human genome project that uses restriction enzymes to break the DNA of each chromosome into a number of identifiable fragments, which are then cloned, is the:
    A. genetic mapping stage.
    B. DNA-sequencing stage.
    C. physical mapping stage.
    D. whole-genome shotgun method.
    E. assembly stage.

49. Which one of the following did Celera Genomics pioneer as part of the Human Genome Project?
    A. genetic mapping stage
    B. DNA-sequencing stage
    C. physical mapping stage
    D. whole-genome shotgun method
    E. assembly stage

50. True or False?   The functions of all human genes <u>have</u> been determined for the human genome.

51. True or False?   The human genome project used DNA from <u>one human</u>.

52. Research on lung cancer has revealed that the treatment of some types of cancer may be tailored to the specific _____ of each patient.

PROTEOMICS

53. To better understand the structures and functions of cells and organisms, scientists are also studying proteomics because in organisms:

    A. proteins outnumber genes and proteins carry out cell activities.

    B. proteins outnumber genes and genes carry out cell activities.

    C. genes outnumber proteins and proteins carry out cell activities.

    D. genes outnumber proteins and genes carry out cell activities.

54. True or False?   The study of genes and the proteins they encode is helping biologists understand how all of these parts interact within an organism.

55. The systematic study of full protein sets that a genome encodes is called _____.

Human Gene Therapy
TREATING SEVERE COMBINED IMMUNODEFICIENCY

56. Which one of the following statements is *false?*

    A. The goal of gene therapy is to replace a mutant version of a gene with a properly functioning one within a living person.

    B. It was not until 2000 that the first scientifically strong evidence of effective gene therapy was reported.

    C. Safe and effective gene therapy is now widely used in medicine.

    D. Severe combined immunodeficiency (SCID) is a fatal inherited disease caused by a single defective gene.

57. True or False?   Human gene therapy is a recombinant DNA procedure that alters a living genome by introducing natural genetic material.

58. One of the prime targets for gene therapy are bone _____ cells that give rise to all the cells of the blood and immune system.

Safety and Ethical Issues
THE CONTROVERSY OVER GENETICALLY MODIFIED FOODS

59. Which one of the following statements is *false?*

    A. The European Union suspended the introduction of new GM crops.

    B. GM strains account for a significant percentage of several agricultural crops.

    C. In the United States, labeling of GM foods is now being debated but has not yet become law.

    D. Lawn and crop grasses commonly exchange genes with wild relatives via pollen transfer.

    E. In the United States today, most public concern centers not on genetically modified (GM) foods but on recombinant microbes.

60. True or False?   The U.S. National Academy of Sciences released a study finding <u>no scientific evidence</u> that transgenic crops pose any special health or environmental risks.

ETHICAL QUESTIONS RAISED BY DNA TECHNOLOGY

61. With respect to the ethical issues raised by genetic engineering, the authors argue that:
    A. there really isn't much to be concerned about.
    B. the issues are so troubling that current research should be stopped.
    C. there are serious societal issues that need to be addressed.
    D. the scientific community will find answers to these concerns.
    E. scientists have not been concerned about these issues.

62. True or False?   Genetic engineering of gametes (sperm or ova) and zygotes in humans <u>has not</u> been attempted.

Evolution Connection: Genomes Hold Clues to Evolution

63. Research on the genetics of organisms at all levels of biological organization suggests that:
    A. organisms are not as interrelated as we think.
    B. life does not have unifying principles.
    C. life at all levels is interrelated.
    D. the genetics of yeast and humans are fundamentally different systems.
    E. although primitive forms of life are similar, they share little with multicellular eukaryotes.

64. True or False?   The DNA sequences determined to date <u>do not</u> confirm the evolutionary connections between distantly related organisms.

65. True or False?   Some yeast genes can substitute for similar genes in <u>human</u> cells.

Word Roots

liga = tied (DNA ligase: the enzyme that permanently "pastes" together DNA fragments)

telo = an end (telomeres: the repetitive DNA at chromosome ends)

Key Terms

biotechnology
DNA fingerprinting
DNA ligase
DNA technology
forensics
gel electrophoresis
gene cloning
genetic marker

genetically modified (GM) organism
genomic library
genomics
human gene therapy
nucleic acid probe
plasmid

polymerase chain reaction (PCR)
proteomics
recombinant DNA
recombinant DNA technology
repetitive DNA
restriction enzyme

restriction fragments
STR analysis
transgenic organism
vaccine
vector

Crossword Puzzle

Use the Key Terms list from this chapter to fill in the crossword puzzle.

ACROSS

1. another term for a genetically modified organism
3. abbreviation for a technique that quickly and precisely copies a segment of DNA
4. the cutting tool for making recombinant DNA
6. much of the DNA between genes
7. a harmless variant or derivative of a pathogen used to prevent disease
9. the study of full protein sets encoded by genomes
10. the alternation of the genes of a person afflicted with a genetic disease
13. the entire collection of cloned DNA fragments from a shotgun experiment
15. methods for studying and manipulating genetic material
16. the production of multiple copies of a gene
17. pieces of DNA produced by restriction enzymes
18. the scientific analysis of evidence for legal investigations
19. the enzyme that permanently "pastes" together DNA fragments

20. a specific pattern of electrophoresis bands that are of forensic use
21. abbreviation for the type of analysis that compares repetitive DNA sequences
23. the study of whole sets of genes and their interactions
24. a type of technology that combines genes from different sources

DOWN

2. in DNA technology, a labeled single-stranded nucleic acid molecule used to find a specific gene
5. the role of a plasmid when it carries extra genes to another cell
8. a host that carries recombinant DNA
11. any DNA segment that varies from person to person
12. a method for sorting macromolecules primarily on the basis of their size and electrical charge
14. the use of organisms to perform practical tasks
22. a small ring of DNA separate from the chromosome(s)

CHAPTER 13

How Populations Evolve

Studying Advice

a. A thorough understanding of evolution is necessary to understand biology. This chapter lays the foundation for the common evolutionary theme of the other chapters. It is largely conceptual with relatively few new terms. It is a chapter worth mastering.

b. The Hardy-Weinberg formula may at first seem intimidating. However, spend the time to understand how it can be used to make predictions about the chances of inheriting disease. Several questions are included below to quiz your comprehension.

Student Media

Activities

The Voyage of the *Beagle*: Darwin's Trip Around the World
Darwin and the Galápagos Islands
Reconstructing Forelimbs
Genetic Variation from Sexual Recombination
Causes of Microevolution

Case Studies in the Process of Science

What Are the Patterns of Antibiotic Resistance?
How Do Environmental Changes Affect a Population of Leafhoppers?
How Can Frequency of Alleles Be Calculated?

MP3 Tutor

Natural Selection

Videos

Galápagos Islands Overview

Discovery Channel Video Clip: Charles Darwin

Grand Canyon

Galápagos Marine Iguana

Galápagos Sea Lion

Galápagos Tortoise

You Decide

Is Ephedra Safe and Effective?

Organizing Tables

Define and compare the pairs of concepts in the following.

TABLE 13.1		
	Genetic Drift	**Gene Flow**
Mechanism What is happening?		
Consequence How does this affect the populations?		
	Founder Effect	**Bottleneck Effect**
Mechanism What is happening?		
Consequence How does this affect the populations?		

Compare the three types of selection in the following.

TABLE 13.2

	Directional Selection	Disruptive Selection	Stabilizing Selection
What members of the population are being favored?			
How does the population curve change as a result of this selective force?			

Content Quiz

Directions: Identify the *one* best answer for the multiple-choice questions. For true/false questions, determine if the statement is true or false. If false, change the underlined word(s) to make the statement true. Finally, add the correct word(s) to the fill-in-the-blank questions to make the statements true.

Biology and Society: Persistent Pests

1. The World Health Organization's effort to spray mosquitoes' habitat with DDT in tropical regions of the world ended because:
   A. it was very successful and no longer needed.
   B. it killed the mosquitoes but also the animals that fed on them.
   C. the DDT accumulated in the ecosystem and caused too many other animals to die.
   D. it encouraged the evolution of new strains of mosquitoes resistant to DDT.
   E. none of the above.

2. Pesticide resistance quickly evolves in insects because:
   A. the pesticide creates new individuals that can resist it.
   B. new species evolve that eat the pesticide.
   C. those that survive produce offspring that inherit the genes for pesticide resistance.
   D. the pesticide changes the DNA and creates resistant individuals.
   E. the insects just get used to the presence of the insecticide.

3. True or False?   After one application of a strong pesticide, the percentage of the population that is resistant to the pesticide will <u>decrease</u>.

4. True or False?   Natural selection <u>creates</u> organisms that can resist pesticides.

5. All life is united by _____.

Charles Darwin and "The Origin of Species"
DARWIN'S CULTURAL AND SCIENTIFIC CONTEXT, DESCENT WITH MODIFICATION

6. Which one of the following people helped set the stage for Darwin by proposing that adaptations evolve as a result of interactions between organisms and their environments?
   A. Buffon
   B. Lyell
   C. Lamarck
   D. Wallace
   E. Anaximander

7. In his book *The Origin of Species*, Charles Darwin developed two main points. These were that:
   A. the world is very old and evolution occurs by natural selection.
   B. evolution occurs slowly and new species form by spontaneous generation.
   C. new species evolve by mutations and new species do not reproduce with old species.
   D. modern species descended from ancestral species and organisms evolve by natural selection.

8. Which one of the following statements about Darwin's voyage on the *Beagle* is *false*?
   A. During the trip, Darwin read and was strongly influenced by the book *Principles of Geology* by Charles Lyell.
   B. The concept of natural selection came quickly to Darwin when he discovered 13 species of finches on the Galápagos.
   C. Darwin thought that most of the animals of the Galápagos resembled species living on the South American mainland.
   D. Darwin collected many South American species of plants and animals while on the trip.

9. In his book *Principles of Geology*, Lyell argued that:
   A. Earth is about 6,000 years old.
   B. Earth is ancient, sculpted by gradual geologic processes that continue today.
   C. fossils could be explained by a single worldwide flood that occurred in the last 10,000 years.
   D. erosion, earthquakes, and other geologic forces are new geologic events and did not occur regularly in the past.

10. Which one of the following people also suggested the idea of natural selection?
    A. Buffon
    B. Lyell
    C. Lamarck
    D. Wallace
    E. Anaximander

11. True or False?   Darwin <u>was</u> the first to suggest that life evolves.

12. True or False?   The Greek philosopher <u>Aristotle</u> suggested that simpler forms of life preceded ones that are more complex.

13. True or False?   <u>Lamarck</u> proposed that by using or not using its body parts, an individual develops certain characteristics, which it passes on to its offspring.

14. True or False?   Lamarck explained evolution as a process of <u>adaptation</u>.

Evidence of Evolution
THE FOSSIL RECORD

15. When we examine the fossil record, we find that:
    A. older fossils are in strata below younger fossils.
    B. amphibians appeared before the first fishes.
    C. transitional fossil forms are missing.
    D. the sediments of each strata all come from a common geologic region.
    E. there are major inconsistencies with molecular and cellular evidence.

16. True or False?   The oldest known fossils are single-celled <u>eukaryotes</u>.

17. A preserved remnant or impression of an organism is a(n) _____.

BIOGEOGRAPHY, COMPARATIVE ANATOMY

18. Which one of the following types of evidence first suggested to Darwin that modern organisms evolved from ancestral forms?
    A. comparative embryology
    B. the fossil record
    C. comparative anatomy
    D. biogeography
    E. molecular biology

19. Which one of the following is most like the concept of homology?
    A. mass producing the Model T car
    B. repairing a car that has been in an accident

C. automakers each producing their own version of a pickup truck

D. replacing worn tires, changing the oil, and tuning up the engine of a car

20. Which one of the following is most like the way that evolution works?
    A. remodeling your snow skis into water skis now that you've moved from Alaska to Florida
    B. building a house by gathering a group of architects and engineers to develop a new design
    C. mixing together chemicals to find a solution to remove graffiti from walls
    D. hiring a chemist to discover the secret formula of Coca-Cola

21. True or False?   Biogeography <u>supports</u> the concept that species were individually placed into suitable environments.

COMPARATIVE EMBRYOLOGY, MOLECULAR BIOLOGY

22. The embryos of vertebrates:
    A. appear most similar at the earliest stages of development.
    B. appear most different at the earliest stages of development.
    C. develop similar structures from different embryonic parts.
    D. are dramatically different throughout development.

23. Evolution predicts that two closely related species, such as humans and chimpanzees, would have:
    A. many similar genes and proteins.
    B. similar anatomical patterns.
    C. common embryological stages.
    D. a common fossil ancestor.
    E. all of the above.

24. True or False?   Molecular biology <u>confirmed</u> the fossil record and other evidence supporting Darwin's view of the interrelatedness of all life.

25. Similar genes in two species suggest that the genes have been inherited from a(n) _____.

Natural Selection
DARWIN'S THEORY OF NATURAL SELECTION

26. Which one of the following is *not* an assumption of natural selection?
    A. Populations typically produce more offspring than can survive.
    B. Resources to support a population are unlimited.
    C. Individuals in a population vary.
    D. Many traits are inherited.

27. Darwin suggested that the animals on the Galápagos Islands had become new species because they:

    A. were all much larger than their ancestors.

    B. were reproducing very quickly.

    C. ate foods similar to those eaten by their ancestors.

    D. were adapting to new, local environments.

28. True or False?   The different beaks of the many Galápagos finch species <u>are homologous</u>, because they are variations on the ancestral body plan.

29. True or False?   The most fit individuals tend to leave the <u>fewest</u> fertile offspring.

30. Darwin thought that the Galápagos Islands were first colonized by animals from _____.

31. According to natural selection, a population's inherent variability is screened by the _____.

NATURAL SELECTION IN ACTION

32. Instead of being a creative mechanism, natural selection is really more a process of:

    A. editing.

    B. remixing.

    C. integrating.

    D. converging.

    E. directing.

33. True or False?   An adaptation in one environment may <u>be useless or harmful</u> in different circumstances.

34. True or False?   Doctors have documented an <u>increase</u> in drug-resistant strains of HIV.

The Modern Synthesis: Darwinism Meets Genetics
POPULATIONS AS THE UNITS OF EVOLUTION

35. Which one of the following is the smallest unit of evolution?

    A. a cell

    B. an individual organism

    C. a population

    D. a species

    E. an ecosystem

36. True or False?   Members of separate populations of a species <u>cannot</u> interbreed.

37. True or False?   <u>Individuals</u> evolve.

38. The field of _____ emphasizes the genetic variation within populations and tracks the genetic composition of populations over time.

39. The fusion of genetics with evolutionary biology has become known as the _____.

GENETIC VARIATION IN POPULATIONS

40. Which one of the following processes is *not* very significant in the evolution of most animal and plant species?

    A. mutation

    B. sexual recombination

    C. meiosis

    D. random fertilization

41. True or False?   The <u>phenotype</u> of an organism results from the combined influence of genotype and the environment.

42. The ABO blood groups of humans are an example of a(n) _____ characteristic in which several morphs are common in a population.

ANALYZING GENE POOLS, POPULATION GENETICS AND HEALTH SCIENCE

Matching: Match each part of the Hardy-Weinberg formula, $p^2 + 2pq + q^2 = 1$, to what it represents.

_____ 43. p^2

_____ 44. $2pq$

_____ 45. q^2

_____ 46. 1

A. the frequency of the homozygous recessive genotype

B. the total of the frequency of all genotypes

C. the frequency of the homozygous dominant genotype

D. the frequency of the heterozygous genotype

Matching: Match the term or phrase on the left to its best description on the right.

_____ 47. population

_____ 48. population genetics

_____ 49. modern synthesis

_____ 50. gene pool

_____ 51. polymorphic

A. when two or more morphs are noticeably present

B. the smallest biological unit that evolves

C. all the alleles in all individuals in a population

D. a field that studies genetic variation in a population

E. the combination of Darwin's and Mendel's work

52. If 1% of a population is homozygous recessive for a trait with two alleles, then:

    A. 63% are homozygous dominant.

    B. 49% are homozygous dominant.

    C. 18% are heterozygous.

    D. 25% are heterozygous.

    E. 81% are heterozygous.

MICROEVOLUTION AS CHANGE IN A GENE POOL

53. In a non-evolving population, from generation to generation the frequency of alleles:

    A. changes, but the frequency of the genotypes stays the same.

    B. and the frequency of the genotypes stay the same.

    C. and the frequency of the genotypes change.

    D. stays the same, but the frequency of genotypes changes.

54. A non-evolving population in genetic equilibrium is also called the _____ equilibrium.

55. A generation-to-generation change in a population's frequencies of alleles defines _____.

Mechanisms of Microevolution
GENETIC DRIFT, GENE FLOW, MUTATION

56. Imagine that a global disease spread quickly across Earth killing every human except the people in your biology class. Although the future human population might someday recover, future generations would not represent the current diversity of all humans living today. This loss of diversity and change in the gene pool would be an example of:

    A. the founder effect.

    B. sexual recombination.

    C. gene flow.

    D. the bottleneck effect.

    E. Hardy-Weinberg equilibrium.

57. Another disaster strikes. Your teacher takes your class on a spectacular field trip to study marine mammals, but your ship encounters a terrific storm. Your class and ship's crew land on a remote and uninhabited Pacific island where they live and prosper, never again making contact with other humans. As future generations come and go, the appearance of your island population takes on its own characteristics, representing the process of:

    A. the founder effect.

    B. sexual recombination.

C. gene flow.

D. the bottleneck effect.

E. Hardy-Weinberg equilibrium

58. Which one of the following statements about gene flow is *false*?

    A. Modern transportation and the frequent relocation of people has increased gene flow in the human population.

    B. Gene flow occurs when fertile individuals move between populations.

    C. Populations may gain or lose alleles through gene flow.

    D. Gene flow tends to increase genetic differences between populations.

59. Which one of the following would *not* contribute to gene flow?

    A. Very light seeds are easily blown across several miles in high storm winds.

    B. A river floods over into nearby ponds.

    C. A storm blows a flock of birds out to sea where they land on an uninhabited island.

    D. A giant redwood tree drops its seeds all around the base of its grand trunk.

60. True or False?   Genetic bottlenecks usually <u>increase</u> the genetic variability in a population.

61. True or False?   Some human diseases <u>are more abundant</u> in small populations due to genetic drift.

62. True or False?   Genetic drift and gene flow <u>can contribute to</u> microevolution.

63. True or False?   For any one gene locus, mutation alone <u>has</u> a considerable quantitative effect on a large population in a single generation because it is a rare event.

64. The founder and bottleneck effects are examples of _____, an evolutionary mechanism by which the gene pool of a small population changes due to chance.

65. A _____ is a change in an organism's DNA.

NATURAL SELECTION: A CLOSER LOOK

66. Which one of the following terms best represents the process of evolution?

    A. deliberate

    B. intentional

    C. fixed

    D. directional

    E. responsive

67. Which one of the following organisms has the greatest Darwinian fitness?

    A. a robin that finds more food than any other female robin, but lays no eggs

    B. a mother robin that has six chicks hatch, but only one that lives to reproduce

    C. a mother robin that has two chicks, both of which live to reproduce

    D. the largest mother robin in the entire population, who lays one large egg

    E. a mother robin that lays the most eggs, but none of them hatch

Matching: Match the types of selection on the left to descriptions on the right. One item on the left matches two items on the right.

_____ 68. directional selection

_____ 69. stabilizing selection

_____ 70. disruptive selection

    A. selection favoring two or more contrasting morphs

    B. selection favoring one extreme phenotype

    C. selection favoring a particular trait within a narrow range

    D. the most common type of selection

71. True or False? The Hardy-Weinberg equilibrium demands that all individuals in a population <u>be equal</u> in their ability to survive and reproduce.

72. Human birth weights are kept in the range of 3–4 kilograms by _____ selection.

73. The fossil record indicates that when a population is challenged by a new environmental problem, the most common result is _____.

EVOLUTION CONNECTION: POPULATION GENETICS OF THE SICKLE-CELL ALLELE

74. Which one of the following statements about sickle-cell disease is *false*?

    A. The sickle-cell allele is most common in Africa where the malaria parasite is most common.

    B. People heterozygous for the sickle-cell allele are relatively resistant to malaria.

    C. People heterozygous for the sickle-cell allele are less common than people who are homozygous for the sickle-cell allele.

    D. Only people homozygous for the sickle-cell allele develop the disease.

75. The Hardy-Weinberg formula indicates that in a population with a 20% sickle-cell allele frequency:

    A. the sickle-cell allele benefits more of the population than it hurts by causing sickle-cell disease.

    B. the sickle-cell allele benefits less of the population than it hurts by causing sickle-cell disease.

C. the sickle-cell allele benefits about the same amount of the population that it hurts by causing sickle-cell disease.

D. nearly everyone will develop sickle-cell disease and no one will benefit.

E. everyone will benefit and nobody will develop sickle-cell disease.

76. True or False?   <u>Heterozygous</u> individuals are relatively resistant to malaria.

77. Sickle-cell disease is most common in tropical areas where _____ is a major cause of death.

Word Roots

bio = life; **geo** = the earth (biogeography: the geographic distribution of species)
homo = alike (homology: traits that appear similar due to common ancestry)
micro = small (microevolution: evolution at its smallest scale)
poly = many; **morph** = form (polymorphic: a characteristic of a population in which two or more forms are clearly present)

Key Terms

adaptation
biogeography
bottleneck effect
comparative
  anatomy
comparative
  embryology
directional
  selection

disruptive
  selection
evolution
fitness
fossil record
founder effect
gene flow
gene pool
genetic drift

Hardy-Weinberg
  equilibrium
Hardy-Weinberg
  formula
homology
microevolution
modern synthesis
natural selection
polymorphic

population
population
  genetics
stabilizing
  selection
vestigial organs

Crossword Puzzle

Use the Key Terms list from this chapter to fill in the crossword puzzle.

ACROSS

3. a population in which two or more morphs are clearly present
4. the type of selection that maintains variation for a particular trait within a narrow range
5. remnants of structures that served important functions in an organism's ancestors
6. the mechanism for descent with modification
7. the geographic distribution of species
10. the type of selection that shifts the phenotypic curve of a population in favor of some extreme phenotype
12. genetic exchange with another population
13. a population's increase in the frequency of traits suited to the environment
14. all of the alleles in all of the individuals in a population
15. the contribution an individual makes to the gene pool of the next generation relative to the contributions of other individuals
16. a trait that appears similar due to common ancestry
17. genetic change in a population or species over generations
18. the comparison of body structures between different species
19. the type of selection that can lead to a balance of two or more different morphs
20. the fusion of genetics with evolutionary biology
21. the ordered sequence of fossils as they appear in rock layers, marking the passage of geologic time
22. genetic drift in a new colony
23. the name of the formula and type of equilibrium that describes a population's gene pool

DOWN

1. evolution at its smallest scale
2. a group of interacting individuals belonging to one species and living in the same geographic area
7. genetic drift due to a drastic reduction in population size
8. the study of genetic variation within a population and over time
9. the comparison of structures that appear during the development of different organisms
11. a change in a gene pool of a small population due to chance

How Biological Diversity Evolves

Studying Advice

a.   Chapter 14 builds heavily upon chapter 13. If, for some reason, you have not studied chapter 13 before reaching this chapter, you should read chapter 13 first.

b.   Chapter 14 addresses some very broad and exciting questions about evolution: Why have major groups of organisms gone extinct? How do new species evolve? How have geological processes such as volcanoes, earthquakes, and continental drift affected the evolution of animals? Many students find this chapter to be one of the most interesting in the book. The new terminology is limited. So enjoy the ideas as you reflect on the major forces that have influenced the evolution of life on Earth!

Student Media

Activities

Mechanisms of Macroevolution

Exploring Speciation on Islands

Polyploid Plants

Paedomorphosis: Morphing Chimps and Humans

The Geologic Time Scale

Classification Schemes

Case Studies in the Process of Science

How Do New Species Arise by Genetic Isolation?

How Is Phylogeny Determined Using Protein Comparisons?

MP3 Tutors

Speciation

Videos

Discovery Channel Video Clip: Mass Extinctions
Lava Flow
Volcanic Eruption

You Decide

Can We Prevent Species Extinction?

Organizing Tables

List with examples the five prezygotic and three postzygotic barriers between species.

TABLE 14.1	
Prezygotic barriers	**Examples**
Postzygotic barriers	**Examples**

Distinguish between allopatric and sympatric speciation, noting examples of each.

TABLE 14.2

	Examples	Is It Common in Plants, Animals, or Both?
Allopatric speciation		
Sympatric speciation		

Distinguish between the four main geologic divisions.

TABLE 14.3

Geologic Divisions	Range of Time in Millions of Years Ago	Dominant Forms of Life
Cenozoic		
Mesozoic		
Paleozoic		
Precambrian		

Content Quiz

Directions: Identify the *one* best answer for the multiple-choice questions. For true/false questions, determine if the statement is true or false. If false, change the underlined word(s) to make the statement true. Finally, add the correct word(s) to the fill-in-the-blank questions to make the statements true.

Biology and Society: One Troublesome Species or Two?

1. The West Nile virus is more of a concern on the North American continent than in Europe because:
   A. the mosquito species that spreads the West Nile virus is not found in Europe.
   B. in Europe, the mosquito species that spreads the West Nile virus has evolved mechanisms to kill the virus inside the mosquito before transmitting it to people.
   C. in North America, the mosquito species that spreads the West Nile virus is a blend of two species that remain distinct in Europe and thus do not bite both birds and people there.
   D. in Europe, the mosquito species that spreads the West Nile virus is eaten in very high numbers by a bird species found only in Europe.
   E. None of the choices are correct.

2. True or False?   The mosquito species that spreads the West Nile virus on the North American continent is a <u>hybrid</u> of two mosquito species found in Europe.

3. In North America but not in Europe, the mosquito species that spreads the West Nile virus bites both _____ and _____.

Macroevolution and the Diversity of Life

4. Which one of the following is *not* included in the field of macroevolution?
   A. the multiplication of species
   B. the origin of evolutionary novelty
   C. the explosive diversification following some evolutionary breakthrough
   D. mass extinctions
   E. All of the above are included in the field of macroevolution.

5. True or False?   Biological diversity is generated by <u>linear evolution</u>.

6. Darwin visited the volcanic _____ Islands, named for their giant tortoises.

7. The major changes in the history of life, which include mass extinctions and the origin of evolutionary novelty, are collectively called _____.

The Origin of Species
WHAT IS A SPECIES?

8. According to the "biological species concept," what keeps species separate?
    A. time
    B. differences in diet
    C. natural selection
    D. reproductive barriers
    E. appearance

9. True or False?   All humans belong to <u>the same</u> species.

10. True or False?   Members of a biological species <u>cannot</u> interbreed with members of other species.

11. Because there can be no reproduction, biologists distinguish _____ species mainly by their differences in appearance.

REPRODUCTIVE BARRIERS BETWEEN SPECIES

12. Which one of the following is a postzygotic reproductive barrier?
    A. temporal isolation
    B. habitat isolation
    C. behavioral isolation
    D. hybrid inviability
    E. gamete isolation
    F. mechanical isolation

13. Two closely related species of tree frogs in Illinois cannot be distinguished from each other except for the male's chirping calls, used during the breeding season. This is an example of two species separated by:
    A. temporal isolation.
    B. habitat isolation.
    C. behavioral isolation.
    D. hybrid inviability.
    E. gamete isolation.
    F. mechanical isolation.

14. True or False?   According to the biological species concept, the evolution of <u>reproductive barriers</u> is the key to the origin of new species.

15. A mule, formed by the hybridization of a horse and a male donkey, is an example of _____, a form of reproductive isolation.

MECHANISMS OF SPECIATION

16. Sympatric speciation:
    A. typically occurs over millions of years.
    B. occurs when a population becomes geographically isolated.
    C. is widespread among animals and rare in plants.
    D. can occur in a single generation.
    E. is the mechanism by which most Galápagos species have evolved.

17. In North America, some rare salamanders are triploid, possessing three copies of every chromosome. (They are an all-female species that reproduces by producing triploid eggs that develop without the addition of a sperm nucleus.) It is believed that these salamanders formed when the sperm from one species fertilized a diploid egg of another species. This evolutionary scenario is an example of:
    A. hybrid sterility.
    B. hybrid inviability.
    C. allopatric speciation.
    D. sympatric speciation.

18. True or False?   Most polyploid species arise from <u>one</u> parent species.

19. Most of the plant species we grow for food are the result of _____ speciation.

In the blanks for questions 20–24, write A if the example is of allopatric speciation or S if the example is of sympatric speciation.

_____ 20. A flock of birds is blown by a hurricane onto a remote Pacific island, where the birds remain isolated and become a new species.

_____ 21. A plant seed fails to undergo meiosis, fertilizing itself, and doubling the number of chromosomes in the next generation. These new plants do not interbreed with the parent species.

_____ 22. A long peninsula becomes a string of islands, separating several ant populations and evolving new ant species.

_____ 23. A glacier advances through the midwestern United States, splitting the populations of rabbits into eastern and western species.

_____ 24. Two species of moths hybridize, producing polyploid individuals that remain reproductively isolated and form a new polyploid species.

25. The concept of punctuated equilibrium suggests that a species:
    A. evolves quickly when it first forms, and then changes very little for a long time.
    B. stays very similar when it first forms, but then undergoes drastic change just before becoming another species.
    C. stays the same when it first evolves, then changes a lot and then not much at all, repeatedly, until it becomes another species.
    D. changes a lot when it first evolves, then it goes extinct.

26. The somewhat sudden switch in the music industry from vinyl records to compact discs is most like:
    A. sympatric speciation.
    B. allopatric speciation.
    C. punctuated equilibrium.
    D. microevolution.

27. True or False?   A "sudden" geologic appearance of a species may actually <u>be 50,000–100,000 years</u>.

28. True or False?   Once a species forms, species in a stable environment may experience <u>much</u> noticeable change.

The Evolution of Biological Novelty
ADAPTATION OF OLD STRUCTURES FOR NEW FUNCTIONS

29. Which one of the following terms is closest in meaning to the term <u>exaptation</u>?
    A. building
    B. remodeling
    C. destroying
    D. repairing
    E. anticipating

30. True or False?   Lightweight bones <u>could have</u> evolved in the ancestors of birds in anticipation of the evolution of flight in birds.

31. A structure that evolved in one context but conveyed advantages for other functions is a(n) _____.

32. Which one of the following is *not* a paedomorphic trait of humans?
    A. large brain
    B. rounded skull
    C. large teeth
    D. flat face
    E. small jaws

33. The evolution of dramatically different species by the process of paedomorphosis is most like:
    A. ordering a birthday cake without frosting.
    B. blending a new flavor of wine by mixing wine from red and white grapes.
    C. adding extra luxuries to an automobile to make a special edition.
    D. repairing a truck after a collision.
    E. creating a tetraploid plant by hybridizing two diploid species.

34. Examine the transformation grids in Figure 14.15. Which one of the following shows the *greatest* change in the development of the chimpanzee into an adult?
    A. top of the skull
    B. eye region
    C. region around the eyes
    D. region around the mouth (upper and lower jaw)
    E. bottom rear of the skull

35. True or False?   Slight genetic changes in <u>key genes controlling development</u> can dramatically change the appearance of an animal.

36. The retention of juvenile traits in the adult, or _____, has occurred in salamanders and humans.

Earth History and Macroevolution
GEOLOGIC TIME AND THE FOSSIL RECORD

37. Pulling away many layers of wallpaper as a person remodels an old home is most like:
    A. the way that the continents formed.
    B. the distribution of fossils within a sedimentary layer.
    C. exploring layers of sedimentary rock.
    D. the effects of earthquakes on the continents.
    E. radiometric dating.

38. Which one of the following commonly marks the boundaries of geologic eras?
    A. evidence of extensive earthquake activity
    B. sudden changes in ocean levels
    C. little change in biological diversity
    D. mass extinctions
    E. alignment of certain stars

39. Which one of the following is the correct sequence of the four major eras, from oldest to most recent?
    A. Paleozoic, Mesozoic, Precambrian, Cenozoic
    B. Precambrian, Paleozoic, Mesozoic, Cenozoic
    C. Cenozoic, Mesozoic, Paleozoic, Precambrian
    D. Cenozoic, Precambrian, Paleozoic, Mesozoic
    E. Precambrian, Mesozoic, Paleozoic, Cenozoic

40. Examine Figure 14.17 and read the figure legend to understand the half-life of an element. If an element has a half-life of 20,000 years, and we start out with 1 kilogram of the element, how much of the element will be present after 40,000 years?
    A. 4 kilograms
    B. 2 kilograms
    C. 0.5 kilograms
    D. 0.25 kilograms
    E. 0.10 kilograms

41. True or False?   Examining the fossils in layers of sedimentary rock reveals the <u>absolute</u> age of the layers.

42. The rate of radioactive decay of specific isotopes can be used to date fossils in a process called _____.

PLATE TECTONICS AND MACROEVOLUTION

43. Which one of the following did *not* occur as a result of the formation of Pangaea?
    A. Ocean levels were lowered.
    B. Shallow marine communities dried out.
    C. The total amount of shoreline was increased.
    D. Populations of organisms that had evolved in isolation were brought together.
    E. The continental interior increased in size.

44. If we compared the fossils of vertebrates found in Africa and South America, we would expect fossils found about 200 million years ago to be:

    A. similar but those from animals 50 million years ago to be quite different.

    B. different, but those from animals 50 million years ago to be quite similar.

    C. similar, just like fossils from animals 50 million years ago.

    D. different, just like fossils from animals 50 million years ago.

45. True or False?   The breakup of Pangaea probably caused <u>sympatric</u> speciation in many animals.

46. When Pangaea broke up, the continents moved apart because of _____.

MASS EXTINCTIONS AND EXPLOSIVE DIVERSIFICATION OF LIFE

47. Which of the following is considered to be a contributing factor to the demise of the dinosaurs 65 million years ago?

    A. increased volcanic activity

    B. a cooled global climate

    C. an asteroid or comet striking Earth

    D. All of the above are contributing factors.

48. Throughout the history of life on Earth, mass extinctions have been followed by periods of:

    A. little life.

    B. recovery of many of the same species.

    C. no life at all.

    D. an explosion of new diversity.

49. True or False?   Over the last 600 million years, there have been <u>two</u> distinct periods of mass extinctions.

Classifying the Diversity of Life
SOME BASICS OF TAXONOMY

50. Which one of the following is the correct sequence of taxonomic groups moving from most general to most specific?

    A. domain, kingdom, phylum, class, order, family, genus, and species

    B. kingdom, domain, order, class, family, phylum, genus, and species

    C. phylum, kingdom, domain, class, order, genus, family, and species

    D. domain, kingdom, phylum, order, class, family, genus, and species

    E. class, kingdom, phylum, order, genus, family, species, and domain

51. Which one of the following is the correct way to write the scientific name for humans?
    A. *Homo Sapiens*
    B. homo Sapiens
    C. *homo sapiens*
    D. Homo sapiens
    E. *Homo sapiens*

52. The hierarchical system used to classify life is most like:
    A. the many ranks in the military.
    B. the many names for colleges in the United States.
    C. the different types of fruits and vegetables in a grocery store.
    D. the first, last, and middle names that people use.

53. True or False?   The first part of the scientific name of several species in the same genus will <u>all be the same.</u>

54. Carolus Linnaeus suggested a system that gives each species a two-part name, or _____.

CLASSIFICATION AND PHYLOGENY

55. Homologous structures are related to each other in the same way as:
    A. two different animal figures are to each other, when both are carved from bars of soap.
    B. two buildings are to each other, when one is made from brick and the other from steel.
    C. pens are to pencils.
    D. the brain is to the stomach.
    E. water is to fish.

56. Organisms that are very closely related are more likely to have:
    A. different genes and few homologous structures.
    B. different genes and many homologous structures.
    C. similar genes but few homologous structures.
    D. similar genes and many homologous structures.

57. Examine Figure 14.22 showing the classification of four carnivores. Which of the following pairs of genera are most closely related to each other?
    A. *Canis* and *Mephitis*
    B. *Canis* and *Lutra*
    C. *Panthera* and *Mephitis*
    D. *Panthera* and *Canis*
    E. *Mephitis* and *Lutra*

58. Examine the evolutionary relationships in Figure 14.24. The common ancestor of the group that included just the red kangaroo and the North American beaver had these characteristics:

  A. vertebral column but no hair, no mammary glands, and no gestation.

  B. vertebral column, hair, no mammary glands, and no gestation.

  C. vertebral column, hair, mammary glands, but no gestation.

  D. vertebral column, hair, mammary glands, and gestation.

  E. vertebral column, hair, mammary glands, and egg laying.

59. True or False?   Convergent evolution often produces structures that are homologous but not analogous.

60. Homologous structures have evolved from _____ in a common ancestor.

61. Structures that are _____ to each other are similar in function.

62. Phylogenetic trees are based upon _____ relationships.

63. Very complex systems are less likely to be the result of _____ evolution and more likely to be the result of _____.

64. One of the most widely used methods in systematics, _____, uses computers to analyze data to identify clades with unique homologies.

ARRANGING LIFE INTO KINGDOMS: A WORK IN PROGRESS

65. In the three-domain system, the new domains Bacteria and Archaea are:

  A. subdivisions of the kingdom Protista.

  B. subdivisions of the kingdom Fungi.

  C. subdivisions of the kingdom Plantae.

  D. groups that were not discovered until the 1970s.

  E. groups that contain prokaryotic cells.

66. True or False?   Although seaweeds are multicellular, they are commonly classified into the typically unicellular kingdom Plantae.

67. The revision of the five-kingdom system to the three-domain system was largely based upon molecular studies and _____.

Evolution Connection: Just a Theory?

68. In science, a "theory":

    A. is about the same as a hypothesis.

    B. accounts for many facts and attempts to explain many phenomena.

    C. is a strict rule that has no exceptions and is not to be questioned.

    D. is really a type of law, a rule that is widely applied in all fields of science.

    E. is an idea in its earliest stages of development, awaiting experimental evidence to test it.

69. True or False?   Among scientists, debates over evolution are like debates over gravity, in which scientists seek to understand the <u>mechanisms</u> of obvious phenomena.

Word Roots

allo = other; **patric** = country (allopatric: the type of speciation in which a new species forms by geographic isolation)

bi = two; **nom** = name (binomial: a two-part name used to identify a species)

con = together (convergent evolution: the type of evolution in which unrelated organisms evolve structures with similar functions)

ex = beyond (exaptation: structures that evolve in one context and later become adapted for other functions)

macro = large (macroevolution: major biological changes evident in the fossil record)

paedo = child; **morphosis** = shaping (paedomorphosis: the retention in the adult of features that were juvenile in its ancestors)

post = after (postzygotic barriers: the type of barriers that prevent development of a zygote that is a hybrid between species)

pre = before (prezygotic barriers: the type of barriers that prevent the fertilization of the eggs of different species)

sym = together; **patri** = habitat (sympatric: the type of speciation in which a new species forms in the same geographic region)

Key Terms

allopatric
  speciation
analogy
binomial
biological species
  concept
clade
cladistics
class
convergent
  evolution

domain
exaptation
family
genus
geologic time scale
kingdom
macroevolution
order
paedomorphosis
phyla

phylogenetic tree
phylogeny
postzygotic
  barriers
prezygotic barriers
punctuated
  equilibrium
radiometric dating

speciation
species
sympatric
  speciation
systematics
taxonomy
three-domain
  system

Crossword Puzzle

Use the Key Terms list from this chapter to fill in the crossword puzzle.

ACROSS

5. a group of classes
7. the study of the diversity and relationships of organisms, past and present
8. the system of classification that divides all life into three groups
9. the origin of new species
11. the type of systematics that uses computers and homologous structures
14. the evolution of a structure in one context that becomes adapted for other functions
17. the type of concept that defines a species
18. the identification, naming, and classification of species
19. anatomical similarity due to convergent evolution
22. a branching diagram that represents a hypothesis about evolutionary relationships among organisms
24. in classification, the taxonomic category just below genus
25. the first part of a binomial
27. the type of barriers that prevent development of a zygote that is a hybrid between species
28. the broadest category of classification that contains kingdoms
29. a table of historical periods grouped into four eras

DOWN

1. the process by which new species form in spurts of rapid change followed by periods of slow speciation
2. major biological changes evident in the fossil record
3. the retention of juvenile body features in the adult stage
4. the type of barriers that prevent the fertilization of the egg of different species
6. the use of half-lives to identify the age of fossils
10. a two-part, Latinized name of a species; for example, *Homo sapiens*
12. a group of orders
13. a group of genera
15. the type of evolution in which unrelated organisms evolve structures with similar functions
16. the evolutionary history of a species
19. the type of speciation in which a new species forms by geographical isolation
20. a group of families
21. the type of speciation in which a new species forms without geographic isolation
23. a group of phyla
26. an ancestral species and all of its descendants

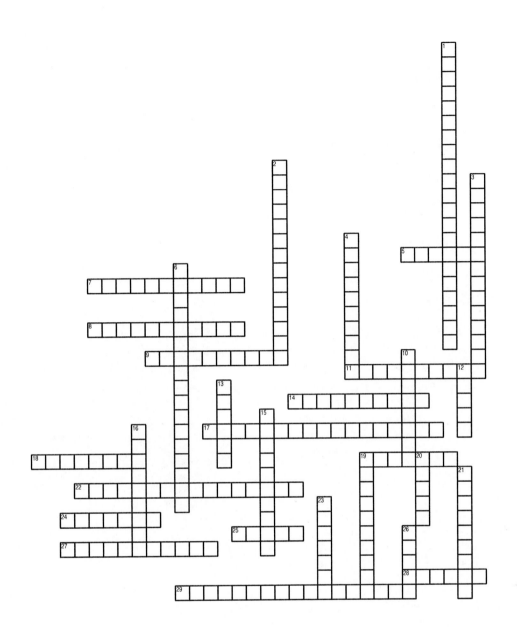

The Evolution
of Microbial Life

Studying Advice

a. Chapter 15 is the first of several survey chapters (chapters 15, 16, and 17) that introduce quite a few names. Many students find that making note cards, with the group name on one side and its characteristics on the back, helps them organize the groups and learn them for exams. As you quiz yourself, remember to keep track of those questions you miss so that you can review those again. Consider alternating the way you quiz yourself, sometimes starting with the group name and other times starting with the group descriptions.

b. The five organizing tables will help you review some of the key chapter information.

Student Media

Activities

The History of Life
Prokaryotic Cell Structure and Function
Diversity of Prokaryotes

Case Studies in the Process of Science

How Did Life Begin on Early Earth?
What Are the Modes of Nutrition in Prokaryotes?
What Kinds of Protists Are Found in Various Habitats?

MP3 Tutors

The Diversity of Prokaryotes

Videos

Discovery Channel Video Clip: Early Life
Prokaryotic Flagella
Discovery Channel Video Clip: Bacteria
Discovery Channel Video Clip: Tasty Bacteria

Cyanobacteria
Euglena
Euglena motion
Stentor
Stentor Ciliate Movement
Vorticella Cilia
Vorticella Detail
Vorticella Habitat
Diatoms Moving
Various Diatoms
Dinoflagellate
Amoeba
Amoeba Pseudopodia
Plasmodial Slime Mold Streaming
Plasmodial Slime Mold
Volvox Colony
Volvox Daughter
Volvox Flagella
Chlamydomonas

Organizing Tables

Identify the key events that occurred during the periods in Earth's history listed in the table.

TABLE 15.1	
Billions of Years Ago (BYA)	Key Events in Earth's History and the Evolution of Life
4.5	
4.0	
by 3.5	
3.5–2.5	
2.5	
2.2	
About 1.0	
0.54 (540 million years ago)	
0.475 (475 million years ago)	

Describe the four main stages, in order, of the hypothesis for the evolution of life on Earth.

TABLE 15.2	
	Key Events of Stage
Stage 1	
Stage 2	
Stage 3	
Stage 4	

Use Table 15.3 to compare the prokaryotes and eukaryotes.

TABLE 15.3		
	Prokaryotes	Eukaryotes
Domains		
Kingdoms		
Examples of each kingdom		
Is a true nucleus present?	Yes or No	Yes or No
Are other membrane-enclosed organelles present?	Yes or No	Yes or No
Are cell walls present? If yes, are they made of cellulose?	Yes or No	Yes or No

Use Table 15.4 to compare the different types of bacteria.

TABLE 15.4		
Bacterial Cell Shape	Draw the Shape	Example from the Text
Cocci		
Staphylococci		
Streptococci		
Bacilli		No example given in the text
Spirochetes		

Use Table 15.5 to compare the types of nutritional strategies.

TABLE 15.5			
Prokaryotic Group	Use Light Energy to Synthesize Organic Compounds?	Require at Least One Organic Nutrient as a Source of Carbon?	Examples Given in Text
Photoautotrophs	Yes or No	Yes or No	
Chemoautotrophs	Yes or No	Yes or No	
Photoheterotrophs	Yes or No	Yes or No	
Chemoheterotrophs	Yes or No	Yes or No	

Content Quiz

Directions: Identify the *one* best answer for the multiple-choice questions. For true/false questions, determine if the statement is true or false. If false, change the underlined word(s) to make the statement true. Finally, add the correct word(s) to the fill-in-the-blank questions to make the statements true.

Biology and Society: Bioterrorism

1. Which one of the following types of organisms has *not* been used for bioterrorism?
    A. animals
    B. plants
    C. fungi
    D. viruses
    E. All of the above have been used for bioterrorism.

2. True or False?   The United States once had a <u>bioweapons</u> program.

3. During the Middle Ages, the bacterium that causes _____ was used as a biowarfare agent.

Major Episodes in the History of Life

4. Which one of the following is the correct sequence of the evolution of early life?
    A. prokaryotic cells, photosynthetic prokaryotes, prokaryotes using cellular respiration, eukaryotic cells, multicellular eukaryotes
    B. photosynthetic prokaryotes, prokaryotic cells, eukaryotic cells, prokaryotes using cellular respiration, multicellular eukaryotes
    C. eukaryotic cells, photosynthetic eukaryotes, prokaryotes using cellular respiration, prokaryotic cells, multicellular prokaryotes
    D. multicellular eukaryotes, eukaryotic cells, prokaryotic cells, photosynthetic prokaryotes, prokaryotes using cellular respiration
    E. prokaryotic cells, prokaryotes using cellular respiration, photosynthetic prokaryotes, eukaryotic cells, multicellular eukaryotes

5. Which one of the following is *false*?
    A. Multicellularity evolved about 1 billion years ago.
    B. The oldest known eukaryotes are about 3.0 billion years old.
    C. Life was confined to water for nearly 85% of its existence on Earth.
    D. Plants and fungi were the first to colonize land.
    E. Photosynthesis evolved about 2.5 billion years ago.

6. Examine the evolutionary relationships in Figure 15.2. Which one of the following pairs is most closely interrelated?

    A. archaea and plants

    B. archaea and protists

    C. bacteria and protists

    D. fungi and animals

    E. animals and protists

7. True or False?   For almost 2 billion years, <u>eukaryotes</u> lived alone.

8. True or False?   Mitochondria and chloroplasts are descendants of <u>prokaryotes</u>.

9. The addition of great amounts of oxygen into our atmosphere about 2.5 billion years ago favored organisms that could use the process of _____.

The Origin of Life
RESOLVING THE BIOGENESIS PARADOX

10. Biogenesis is the idea that:

    A. life emerges by spontaneous generation.

    B. the first cells ever to evolve arose by spontaneous generation.

    C. it takes a male and female organism to produce new life.

    D. life is cellular.

    E. life gives rise to life.

11. True or False?   The conditions on Earth when life first arose were very <u>similar</u> to conditions today.

12. The idea that life can come from nonliving matter, called _____, was commonly accepted before modern science demonstrated otherwise.

A FOUR-STAGE HYPOTHESIS FOR THE ORIGIN OF LIFE, FROM CHEMICAL EVOLUTION TO DARWINIAN EVOLUTION

13. Which one of the following is the correct sequence of the four stages of the origin of the first cells?

    A. joining of monomers into polymers, abiotic synthesis of small organic molecules, packaging into pre-cells, and origin of self-replicating molecules

    B. origin of self-replicating molecules, joining of monomers into polymers, abiotic synthesis of small organic molecules, and packaging into pre-cells

    C. origin of self-replicating molecules, abiotic synthesis of small organic molecules, joining of monomers into polymers, and packaging into pre-cells

    D. abiotic synthesis of small organic molecules, joining of monomers into polymers, origin of self-replicating molecules, and packaging into pre-cells

14. Which of the following molecules were produced by experiments simulating conditions on primitive Earth?

    A. nucleotides

    B. sugars and lipids

    C. ATP

    D. all 20 amino acids

    E. all of the above

15. True or False?   It appears that RNA evolved <u>before</u> DNA.

16. True or False?   Natural selection <u>would not have</u> operated on the first pre-cells to have formed.

17. The early replication of RNA may have been catalyzed by _____.

Prokaryotes
THEY'RE EVERYWHERE!

18. Which one of the following statements about prokaryotes is *false*?

    A. Some prokaryotes decompose organic matter and thus help recycle essential nutrients.

    B. Prokaryotes live in many places on Earth where eukaryotes cannot survive.

    C. Some prokaryotes cause serious disease.

    D. Some prokaryotes provide us with vitamins.

    E. Prokaryotes have lived with eukaryotes for about 3.5 billion years.

    F. Some prokaryotes prevent fungal infections in our body.

19. True or False?   Without prokaryotes, eukaryotic life <u>would be dead</u>.

20. True or False?   Without eukaryotes, prokaryotic life <u>would be dead</u>.

21. Tuberculosis, cholera, food poisoning, and many sexually transmissible diseases are caused by _____.

THE TWO MAIN BRANCHES OF PROKARYOTIC EVOLUTION: BACTERIA AND ARCHAEA

22. Bacteria and the archaea both:

    A. live in extreme environments.

    B. occur in about equal numbers in most ecosystems.

    C. have a prokaryotic cellular organization.

    D. produce methane as a waste product of metabolism.

    E. aid in digestion in cattle.

23. True or False?   Some archaea live in <u>aerobic</u> environments and give off methane as a by-product.

24. The _____ are the prokaryotes most closely related to the eukaryotes.

THE STRUCTURE, FUNCTION, AND REPRODUCTION OF PROKARYOTES

Matching: Match the word on the left to its best description on the right.

_____ 25. spirochete   A. chains of cocci bacteria

_____ 26. bacilli   B. clusters of cocci bacteria

_____ 27. cocci   C. rod-shaped bacteria

_____ 28. streptococci   D. dormant bacterial cells

_____ 29. staphylococci   E. spherical bacteria

_____ 30. endospores   F. spiral-shaped bacteria

31. Which one of the following statements about prokaryotes is *false?*

    A. Very few prokaryotic species are motile.

    B. Some prokaryotic species exhibit simple multicellular organization.

    C. Some endospores can remain dormant for centuries and survive boiling water.

    D. Many motile prokaryotes use flagella to move about.

    E. If resources are available, prokaryotic populations can double every 20 minutes.

32. True or False?   Most bacteria have a cell wall exterior to their plasma membrane that is made of the same materials as plant cells.

33. Prokaryotes reproduce by the process of _____.

THE NUTRITIONAL DIVERSITY OF PROKARYOTES

Matching: Match the word on the left to its best description on the right.

_____ 34. photoheterotroph   A. energy source is sunlight; carbon source is CO_2

_____ 35. chemoautotroph

_____ 36. chemoheterotroph   B. energy source is sunlight; carbon source is organic compounds

_____ 37. photoautotroph
    C. energy source is inorganic compounds; carbon source is CO_2

    D. energy source and carbon source are both organic compounds

38. Which of the following nutritional strategies is (are) unique to prokaryotes?

    A. photoautotroph

    B. chemoheterotroph

    C. photoautotroph and chemoheterotroph

    D. chemoautotroph and photoheterotroph

39. True or False?   All plants and algae are chemoautotrophs.

40. All fungi and animals are _____.

THE ECOLOGICAL IMPACT OF PROKARYOTES

41. Most pathogenic bacteria cause disease by:
    A. eating our living flesh.
    B. consuming so much oxygen that our tissues die.
    C. producing nitrogen containing compounds.
    D. producing exotoxins and endotoxins.
    E. living in our cells and disrupting normal cellular processes.

42. The nitrogen that plants use to make proteins and nucleic acids comes from:
    A. photoautotrophs that live in the air.
    B. prokaryotic metabolism in the soil.
    C. bioremediation.
    D. pathogenic bacteria that live inside leaves.
    E. genetically engineered bacteria that we add to the soil.

43. True or False? Most bacteria <u>are</u> pathogenic.

44. Bacteria are used to break down sewage and to clean up oil spills and other toxic environments through the process of _____.

Protists
THE ORIGIN OF EUKARYOTIC CELLS

45. Eukaryotic cells evolved by:
    A. mitosis and meiosis.
    B. inward folds of the plasma membrane and endosymbiosis.
    C. inward folds of the plasma membrane and bioremediation.
    D. endosymbiosis and bioremediation.
    E. mitosis and endosymbiosis.

46. Which one of the following statements about chloroplasts and mitochondria is *false*? Both chloroplasts and mitochondria:
    A. possess DNA.
    B. make some of their enzymes.
    C. reproduce by binary fission.
    D. appear to have evolved by endosymbiosis.
    E. are found in all eukaryotic cells.

47. Examine Figure 15.18. According to the endosymbiosis theory, the inner mitochondrial membrane corresponds to the:

A. outer membrane of a chloroplast.

B. nuclear membrane.

C. original aerobic bacterial membrane.

D. original membrane of a photosynthetic bacterium.

E. endoplasmic reticulum.

48. Examine Figure 15.18 again. According to the endosymbiosis theory, the outer chloroplast membrane corresponds to the:

A. inner mitochondrial membrane.

B. nuclear membrane.

C. plasma membrane of the cell that engulfed it.

D. endoplasmic reticulum.

E. outer membrane of an aerobic bacterium.

49. True or False? The ancestors of mitochondria were probably anaerobic bacteria.

50. Almost all eukaryotes have _____, but only some have chloroplasts.

51. The first eukaryotic cells to evolve were _____.

THE DIVERSITY OF PROTISTS

Matching: Match the group on the left to its best description on the right.

_____ 52. diatoms

_____ 53. flagellates

_____ 54. phytoplankton

_____ 55. dinoflagellates

_____ 56. amoebas

_____ 57. ciliates

_____ 58. plasmodial slime molds

_____ 59. apicomplexans

_____ 60. cellular slime molds

_____ 61. seaweeds

_____ 62. green algae

A. large, multicellular marine algae

B. protozoans covered by cilia

C. grass-green chloroplasts, includes *Volvox*

D. greatly flexible protozoans without permanent locomotory organelles

E. all parasitic protozoans with an apex for penetrating hosts

F. fungal life cycle with a plasmodium feeding stage

G. planktonic algae

H. silica cell wall that consists of two pieces

I. external plates of cellulose and two flagella

J. fungal life cycle with a solitary amoeboid feeding stage

K. protozoans with flagella

Matching: Match the protist group on the left to its characteristic on the right.

_____ 63. flagellates

_____ 64. plasmodial slime molds

_____ 65. ciliates

_____ 66. seaweeds

_____ 67. apicomplexans

_____ 68. dinoflagellates

_____ 69. diatoms

A. mined and used as filtering material or an abrasive

B. their gel-forming substances are used in ice cream and pudding

C. responsible for red tides and massive fish kills

D. can be found among leaf litter on a forest floor

E. *Paramecium* is an example

F. includes the organism that causes malaria

G. includes *Giardia* and trypanosomes that cause sleeping sickness

70. Which of the following is (are) found in nearly all protists?
    A. multicellularity
    B. flagella
    C. chloroplasts
    D. mitochondria
    E. pseudopods
    F. cilia

71. Which one of the following groups consists of organisms that are decomposers?
    A. protozoans
    B. slime molds
    C. unicellular algae
    D. seaweeds

72. Which one of the following groups consists of multicellular photosynthetic protists?
    A. protozoans
    B. slime molds
    C. unicellular algae
    D. seaweeds

73. The cilia on the surface of ciliates move these protists in a way most similar to:
    A. a paddleboat.
    B. a viking ship with one hundred sets of oars.
    C. a sailboat.
    D. a tugboat pushing a ship out to sea.
    E. a submarine using a nuclear engine to turn its propeller.

74. True or False?   The closest relatives of seaweeds are <u>unicellular algae</u>.

75. The _____ are photosynthetic protists that are most related to plants.

76. Seaweeds are classified into three groups based partly on the types of _____ present in their chloroplasts.

Evolution Connection: The Origin of Multicellular Life

77. A major advantage of multicellularity is:
    A. the ability of cells to multiply.
    B. the smaller size of the organism.
    C. faster reproductive cycles.
    D. the ability for cells to specialize.

78. True or False?   Multicellular organisms are <u>fundamentally different from</u> unicellular organisms.

79. The organism _____ is a colonial green algae that shows some of the traits of early multicellular life.

Word Roots

api = the tip (apicomplexans: all parasitic protists named for an apparatus at their apex that is specialized for penetrating host cells)

archae = ancient (archaea: prokaryotic group most closely related to eukaryotes)

bacill = a little stick (bacilli: rod-shaped prokaryotes)

bi = two (binary fission: a form of asexual reproduction in which a single cell divides into two cells of about equal size)

bio = life; **genesis** = origin (biogenesis: the principle that life gives rise to life)

chemo = chemical; **auto** = self (chemoautotrophs: organisms that need only carbon dioxide as a carbon source and extract energy from inorganic substances)

dinos = whirling (dinoflagellates: protists with two flagella that cause them to spin)

endo = inner (endospores: bacterial resting cells)

exo = outside (exotoxins: toxic proteins secreted by bacterial cells)

flagell = a whip (flagellates: protozoa that move by means of one of more flagella)

hetero = different (chemoheterotrophs: organisms that must consume organic molecules for both energy and carbon)

patho = disease (pathogens: organisms that cause disease)

photo = light (photoautotrophs: photosynthetic organisms, including the cyanobacteria)

planktos = wandering (plankton: communities of mostly microscopic organisms that drift or swim near the water surface)

protos = first; **zoan** = animal (protozoan: early protists that ingested food)

pseudo = false; **pod** = foot (pseudopodia: temporary extensions of a cell used for locomotion and/or feeding)

spiro = spiral (spirochetes: spiral-shaped bacteria)

sym = together (endosymbiosis: the process whereby mitochondria and chloroplasts exist as associations between prokaryotic cells living within larger prokaryotic cells)

Key Terms

algae
amoebas
apicomplexans
archaea
bacilli
bacteria
binary fission
biogenesis
bioremediation
cellular slime
 molds

chemoautotrophs
chemoheterotrophs
ciliates
cocci
diatoms
dinoflagellates
endospores
endosymbiosis
endotoxins
eukaryote

exotoxins
flagellates
forams
green algae
pathogens
photoautotrophs
photoheterotrophs
plankton
plasmodial slime
 mold

prokaryote
protists
protozoans
pseudopodia
ribozymes
seaweeds
spirochetes
spontaneous
 generation
symbiosis

Crossword Puzzle

Use the Key Terms list from this chapter to fill in the crossword puzzle.

ACROSS

1. bacterial resting cell

4. toxic protein secreted by bacterial cells

5. protozoans related to amoebas but with calcium carbonate shells

6. protozoans that move by means of one or more flagella

7. the first type of cell that lacks a true nucleus

8. the process that generated mitochondria and chloroplasts

9. organism with solitary amoeboid cells that can form a sluglike colony

11. type of protist characterized by great flexibility and the presence of pseudopodia

14. spherical species of bacteria

15. photosynthetic, plantlike protist

16. protists including Volvox with grass-green chloroplasts

20. organism that needs only carbon dioxide as a carbon source and extracts energy from inorganic substances

21. photosynthetic organism including the cyanobacteria

22. type of cell division in which each daughter cell receives a copy of the single parental chromosome

23. the use of organisms to remove pollutants from water, air, and soil

25. algae and other organisms, mostly microscopic, that drift passively in ponds, lakes, and oceans

26. one of two prokaryotic domains, the other being the bacteria

27. one of two prokaryotic domains, the other being the archaea

29. one of a group of parasitic protozoans, some of which cause human diseases

30. the first eukaryote to evolve from prokaryotic ancestors

32. the principle that life gives rise to life

33. communities of mostly microscopic organisms that drift or swim near the water surface

34. protist with a glassy silica containing cell wall

35. large, spiral-shaped prokaryotic cell

36. a close association between organisms of two or more species

37. large, multicellular marine alga

DOWN

2. protist with two flagella

3. organism that must consume organic molecules for both energy and carbon

4. cells with true nuclei and many other organelles

10. life emerging from inanimate material

12. protist that has a plasmodium feeding stage

13. disease-causing organism

17. toxic chemical component in the cell walls of certain bacteria

18. cellular extensions of amoeboid cells used in moving and feeding

19. organism that obtains energy from sunlight and carbon from organic sources

24. rod-shaped prokaryotic cells

28. type of protozoan that moves by means of cilia

31. an enzymatic RNA molecule that catalyzes reactions during RNA splicing

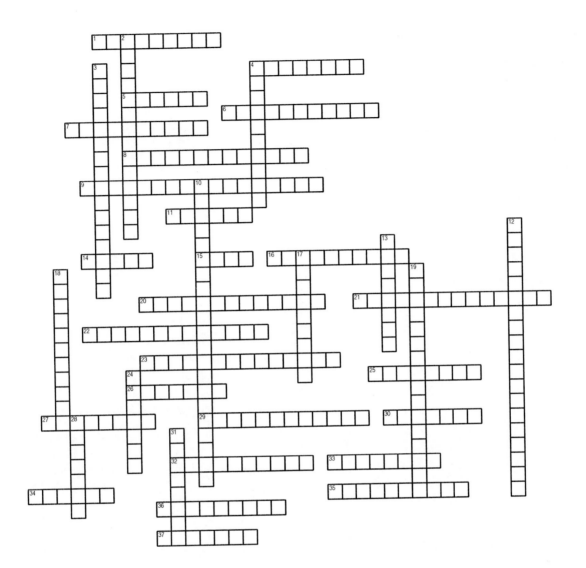

CHAPTER 16

Plants, Fungi, and the Move onto Land

Studying Advice

a. The first three major chapter sections (through "Highlights of Plant Evolution") set the stage for the rest of the chapter. Read these first sections carefully, paying special attention to adaptations to life on land. The next four chapter sections ("Bryophytes" through "Angiosperms") survey these groups in the order they evolved. The adaptations described in the first part of the chapter are referred to regularly. The chapter ends with a brief survey of fungi.

b. If you have ever wondered why ferns, trees, mosses, and mushrooms look so different, this chapter will be interesting. Maybe you have suffered from athlete's foot, ringworm, or a yeast infection. Plants and fungi are all around us. This interesting chapter is a starting point for appreciating their diverse structures and roles in our world.

Student Media

Activities

Terrestrial Adaptations of Plants

Highlights of Plant Evolution

Moss Life Cycle

Fern Life Cycle

Pine Life Cycle

Angiosperm Life Cycle

Madagascar and the Biodiversity Crisis

Fungal Reproduction and Nutrition

Case Studies in the Process of Science

What Are the Different Stages of a Fern Life Cycle?

How Are Trees Identified by Their Leaves?

How Does the Fungus *Pilobolus* Succeed as a Decomposer?

Videos

Discovery Channel Video Clip: Plant Pollination

Flower Blooming (time-lapse)

Bat Pollinating Agave Plant

Bee Pollinating

Flowering Plant Life Cycle (time-lapse)

Discovery Channel Video Clip: Fungi

Earthworm Locomotion

Allomyces Zoospore Release

Phlyctochytrium Zoospore Release

Water Mold Oogonium

Water Mold Zoospores

Flapping Geese

Soaring Hawk

Swans Taking Flight

Organizing Tables

Describe the symbiotic relationships noted in the table below. Two cells are already filled in to help you.

TABLE 16.1		
Type of Mutualistic Relationship	Organisms Involved?	Each Organism's Benefit in the Mutualistic Relationship?
Mycorrhizae		
Plant pollination		The pollinator gets nectar. Pollen is transferred between plants of the same species.
Lichens		
Endosymbiotic origin of mitochondria and chloroplasts (chapter 15)	Mitochondria, chloroplasts, and early eukaryotic cells	

Plant evolution. Indicate the living group that best represents each stage in the evolution of plants. Circle Yes or No to indicate the specific traits found in each group.

TABLE 16.2				
Four Stages of Plant Evolution	Living Example	Gametangia	Vascular Tissue	Seeds Present and Enclosed in Special Chamber
First land plants		Yes or No	Yes or No	No seeds
Diversification of vascular plants		Yes or No	Yes or No	No seeds
Origin of seeds		Yes or No	Yes or No	Yes or No
Flowering plants		Yes or No	Yes or No	Yes or No

Content Quiz

Directions: Identify the *one* best answer for the multiple-choice questions. For true/false questions, determine if the statement is true or false. If false, change the underlined word(s) to make the statement true. Finally, add the correct word(s) to the fill-in-the-blank questions to make the statements true.

Biology and Society: Will the Blight End the Chestnut?

1. The American chestnut trees are seriously threatened because of:
   A. overharvesting by the timber industry during World War II.
   B. competition from Chinese elm trees introduced accidentally in the 1920s.
   C. an Asian fungus introduced into the United States about 1900.
   D. the overharvest of American chestnuts for food by early pioneers.
   E. their inability to tolerate global warming.

2. True or False?   The American chestnut trees were destroyed by a <u>fungus</u>.

3. American chestnuts today are unable to _____ because of the blight fungus.

Colonizing Land
TERRESTRIAL ADAPTATIONS OF PLANTS

4. The authors note that water lilies are like whales because both:
    A. need to come to the surface of the water to breathe.
    B. have very smooth surfaces.
    C. evolved from terrestrial ancestors.
    D. rely upon photosynthesis.
    E. are not found near the polar regions of Earth.

5. Plants are:
    A. unicellular prokaryotes that make organic molecules by aerobic respiration.
    B. unicellular eukaryotes that make organic molecules by photosynthesis.
    C. multicellular prokaryotes that make organic molecules by photosynthesis.
    D. multicellular eukaryotes that make organic molecules by aerobic respiration.
    E. multicellular eukaryotes that make organic molecules by photosynthesis.

6. Which of the following distinguishes plants from large algae?
    A. Plants use photosynthesis and algae do not.
    B. Plants are eukaryotes and algae are prokaryotes.
    C. Plants are multicellular and algae are unicellular.
    D. Plants have terrestrial adaptations and algae do not.
    E. Plants require oxygen and algae do not.

7. The fungus in mycorrhizae gives the plant:
    A. water and essential minerals, and in turn the fungus receives sugars from the plant.
    B. oxygen, and in turn the fungus receives water and sugars from the plant.
    C. carbon dioxide and carbon, and in turn the fungus receives sugars from the plant.
    D. sugars, and in turn the fungus receives water and essential minerals from the plant.
    E. sugars and essential minerals, and in turn the fungus receives water from the plant.

8. Vascular tissues in plants function most like:
    A. highways for distributing resources within a country.
    B. the Internet for quick and reliable communication between parts of our country.
    C. the steel infrastructure of a high-rise office building.
    D. the extensive air duct system for moving air within a high-rise office building.
    E. an extensive sewage system draining the sewage systems of many apartments.

9. Lignin is to a plant cell as:
    A. muscle is to our bodies.
    B. calcium is to our bones.
    C. telephones are to society.
    D. blood is to our bodies.
    E. hemoglobin is to our blood.

10. The gametangia of plants function most like:
    A. the plastic wrapper around a loaf of bread.
    B. the circulatory system of a mouse.
    C. windows that open up in an apartment.
    D. a can opener.
    E. paddles in a canoe.

11. In most plants, fertilization:
    A. and development occur outside the female parent plant.
    B. occurs within, but development occurs outside the female parent plant.
    C. occurs outside, but development occurs within the female parent plant.
    D. and development occur within the female parent plant.

12. True or False?   Mycorrhizae are found <u>on the roots</u> of some of the oldest plant fossils.

13. Most plants rely upon a symbiotic relationship between their roots and soil fungi, called _____.

14. Carbon dioxide and oxygen move between the atmosphere and the interior of leaves via _____, microscopic holes in the leaf surfaces.

15. Water loss is reduced from the leaf surface by the _____, a waxy coating.

16. Plants use _____ to gather resources below ground and _____ to gather resources from above-ground.

17. The chemical _____ hardens plant cell walls, making them rigid.

THE ORIGIN OF PLANTS FROM GREEN ALGAE

18. The first terrestrial plants likely evolved from:
    A. unicellular algae through a very rapid process.
    B. unicellular algae through a slow and gradual process.
    C. multicellular algae through a very rapid process.
    D. multicellular algae through a slow and gradual process.
    E. multicellular fungi through a very rapid process.

19. True or False?   Gametangia would not have been adaptive in shallow water.

20. The living green algae most closely related to plants are the _____.

Plant Diversity
HIGHLIGHTS OF PLANT EVOLUTION

Matching: Match the group on the left to its best description on the right.

_____ 21. ferns	A.	nonvascular early plants
_____ 22. angiosperms	B.	vascular plants without seeds
_____ 23. gymnosperms	C.	vascular plants with "naked" seeds
_____ 24. bryophytes	D.	vascular plants with flowers

25. True or False?   Most plants alive today are gymnosperms.

26. An embryo packaged along with a store of food within a protective covering defines a(n) _____.

BRYOPHYTES

27. Mosses display two key adaptations to life on land. These adaptations are:
    A. vascular tissue and lignin.
    B. vascular tissue and retention of embryos in the mother plant's gametangium.
    C. lignin and retention of embryos in the mother plant's gametangium.
    D. a waxy cuticle and retention of embryos in the mother plant's gametangium.
    E. a waxy cuticle and vascular tissue.

28. Examine Figure 16.10 showing the life cycle of a moss. Meiosis:
    A. occurs only in the gametophyte stage.
    B. occurs in the sporophyte stage.
    C. occurs in both the gametophyte and sporophyte stages.
    D. does not occur anywhere in the life cycle.

29. True or False?   The cells of the sporophyte are haploid.

30. True or False?   Mosses and other bryophytes are unique in that the <u>sporophyte</u> is the dominant generation.

31. Spores are produced by the _____, and gametes are produced by the _____.

32. A life cycle in which the gametophyte and sporophyte stages produce each other is called _____.

FERNS

33. Ferns are:
    A. nonvascular, seedless plants with flagellated sperm.
    B. vascular, seedless plants with flagellated sperm.
    C. vascular, seedless plants that have pollen.
    D. vascular plants with seeds.
    E. vascular plants with seeds contained in special chambers.

34. True or False?   The greatest amount of coal was deposited during the <u>Carboniferous period</u>.

35. Black sedimentary rock made up of fossilized plant material defines _____.

GYMNOSPERMS

36. Which of the following types of climates favored the evolution of gymnosperms?
    A. wetter and warmer
    B. drier and warmer
    C. drier and colder
    D. wetter and colder

37. Which of the following adaptations to conserve water are found in gymnosperms?
    A. broad leaves, no stomata, and a thick cuticle
    B. broad leaves, stomata in pits, and a thin cuticle
    C. thin leaves, stomata in pits, and a thin cuticle
    D. thin leaves, stomata at the tips of bumps, and a thick cuticle
    E. thin leaves, stomata in pits, and a thick cuticle

38. Compared to ferns, which one of the following was *not* a new terrestrial adaptation of gymnosperms?

    A. vascular tissue

    B. pollen

    C. further reduction of the gametophyte

    D. seeds

39. Examine the three variations on alternation of generation in plants in Figure 16.14. Which one of the figures best represents gymnosperms?

    A. A

    B. B

    C. C

40. True or False?   A conifer is really a <u>sporophyte</u> with tiny <u>gametophytes</u> living in cones.

41. True or False?   Conifers and other gymnosperms <u>have</u> ovaries.

42. Some plant biologists believe that the shift toward the sporophyte as the dominant generation on land was an adaptation to best resist the damaging effects of _____.

43. The much-reduced gametophyte stage that houses cells that will develop into sperm is called _____.

44. In gymnosperms, eggs develop within _____.

ANGIOSPERMS

Matching: Match the flower part on the left to its best description on the right.

_____ 45. sepals	A. a protective chamber containing one or more ovules
_____ 46. petals	B. a stalk bearing an anther
_____ 47. stamen	C. the style, with an ovary at the base and a stigma at its tip
_____ 48. anther	
_____ 49. carpel	D. usually green, they enclose the flower before it opens
_____ 50. ovary	E. the male organ in which pollen grains develop
	F. usually the most attractive part of a flower, attracting pollinators

51. Which one of the following traits is found in angiosperms, but not gymnosperms?
    A. a seed enclosed within an ovary
    B. vascular tissue
    C. an adult gametophyte stage
    D. lignin supporting the cell walls
    E. stomata

52. What are the products of double fertilization?
    A. a zygote and endosperm
    B. fruit and endosperm
    C. a zygote and fruit
    D. an embryo sac and pollen
    E. pollen and fruit

53. True or False?   An angiosperm embryo is nourished by endosperm.

54. True or False?   Most of our food comes from gymnosperms.

55. True or False?   It is the fruit that accounts for the unparalleled success of the angiosperms.

56. The ripened ovary of a flower is the _____.

57. Agriculture is a unique kind of evolutionary relationship between _____ and animals.

PLANT DIVERSITY AS A NONRENEWABLE RESOURCE

58. Which one of the following is the leading cause of forest destruction?
    A. forest fires
    B. hurricanes and tornadoes
    C. human activities
    D. erosion
    E. pollution

59. True or False?   At the current rate of destruction, Earth's tropical rain forests will be gone in 25 years.

60. True or False?   Researchers have investigated the potential medical uses of fewer than 2% of the known plant species.

Fungi
CHARACTERISTICS OF FUNGI

61. Which one of the following statements about fungi is *false*?
    A. Fungi are mostly multicellular.
    B. Fungi are eukaryotes.
    C. Fungi are more closely related to plants than they are to animals.
    D. Fungi decompose organic matter and recycle the nutrients back to the soil.
    E. Some parasitic fungi infect the lungs of humans.

62. Fungi reproduce by:
    A. producing spores by sexual or asexual reproduction.
    B. using flowers.
    C. producing seeds that are fertilized and develop within the female mushroom.
    D. special types of pollen that fertilize each other and then develop underground.
    E. budding or binary fission.

63. The relationship between a mushroom and its hyphae is most like the relationship between:
    A. a fish and a river.
    B. a car and a highway.
    C. a wristwatch and a person's arm.
    D. a fire hydrant and the underground water pipes.
    E. a book and a library.

64. True or False?   The cell walls of fungi are made from <u>cellulose</u>.

65. The structures that account for wide dispersal of fungi are _____.

66. Fungi feed by secreting _____ into the environment and absorbing the digested compounds.

67. The bodies of most fungi are made from minute threads called _____.

68. The feeding network of a fungus, composed of an interwoven mat of hyphae, is called a(n) _____.

THE ECOLOGICAL IMPACT OF FUNGI

69. Which one of the following, if any, is *not* associated with fungi?
    A. mushroom
    B. production of the antibiotic penicillin
    C. athlete's foot
    D. ringworm
    E. vaginal yeast infection
    F. American chestnut blight
    G. truffle
    H. All of the above *are* associated with fungi.

70. True or False?   Animals are <u>more</u> susceptible to parasitic fungi than are plants.

71. Along with bacteria and some invertebrate animals, fungi play an important ecological role as _____ in natural environments.

EVOLUTION CONNECTION: MUTUALISTIC SYMBIOSIS

72. Which one of the following relationships is a type of mutualism?
    A. a thief stealing from a department store
    B. a person picking mushrooms to eat
    C. a hunter shooting ducks
    D. a person buying corn from a farmer
    E. a fungus growing on your shower curtain

73. True or False?   A <u>parasitic</u> relationship is one in which both species in a relationship benefit.

74. Mycorrhizae represent a mutualistic relationship between plant roots and _____.

75. A(n) _____ is composed of fungi and algae that live together in a mutualistic relationship.

Word Roots

angion = a container; **sperma** = seed (angiosperms: flowering plants)

bryo = moss; **phyte** = plant (bryophytes: a group of nonvascular plants that includes the mosses)

endo = inner; **sperm** = seed (endosperm: a source of nutrition for the developing embryo in a seed)

gamet = a wife or husband (gametangia: protective structures where plants produce their gametes)

gymno = naked (gymnosperms: vascular plants such as conifers with seeds not enclosed in specialized chambers)

mutu = reciprocal (mutualism: a symbiotic relationship in which both partners benefit)

myco = fungus; **rhiza** = root (mycorrhizae: symbiotic root-fungus combinations)

sporo = seed (sporophyte: the moss stage with diploid cells)

stoma = mouth (stomata: pores in leaves for gas exchange)

sym = together; **bio** = life (symbiosis: a relationship between two species in which one organism lives in or on another)

Key Terms

absorption
alternation of generations
angiosperms
anther
bryophytes
carpel
charophyceans
conifers
cuticle
double fertilization
embryo sac

endosperm
ferns
flower
fossil fuels
fruit
fungi
gametangia
gametophyte
germinate
gymnosperms
hyphae
lichens
lignin

mosses
mutualism
mycelium
mycorrhizae
ovary
ovules
parasitism
petals
phloem
plant
pollen
roots

seed
sepals
shoots
spores
sporophyte
stamen
stigma
stomata
style
symbiosis
vascular tissue
xylem

Crossword Puzzle

Use the Key Terms list from this chapter to fill in the crossword puzzle.

ACROSS

1. a plant embryo packaged along with a food supply within a protective coat
3. one of a group of seedless vascular plants
5. in a seed plant, the male gametophytes that develop within the anthers of stamens
7. the densely branched network of hyphae in a fungus
8. energy deposits formed from the remains of extinct organisms
9. minute threads of fungi that promote absorptive nutrition
11. typically the most striking parts of a flower, usually important in attracting insects
12. the female gametophyte contained in the ovule of a flowering plant
13. what a seed does when it first begins to grow
14. a chemical that hardens the cell walls of plants
20. a multicellular eukaryote that makes organic molecules by photosynthesis
21. the female reproductive organ of a flower
22. the male organ in a flower in which pollen grains develop
23. protective structures where plants produce their gametes
24. flowering plants
26. the group of vascular plants that bear naked seeds
29. a protective chamber containing one or more ovules
30. the green algal group that is considered to be the closest relative of land plants
31. ecological relationships between organisms of different species in direct contact
32. the most familiar bryophyte group
33. the aerial portion of a plant body, consisting of stems, leaves, and flowers

36. a waxy coating of the leaves and other aerial parts of most plants
37. the portion of the vascular system in plants that transports sugar and other nutrients throughout the plant
38. the stalk portion of the carpel
39. vascular plants that bear naked seeds
40. a symbiotic relationship in which one species benefits while the other species is harmed
41. the moss stage with diploid cells
42. heterotrophic eukaryotes that acquire nutrients by absorption
43. the sticky tip of a flower's carpel, which traps pollen grains
44. the process by which small organic molecules are brought in from the surrounding medium

DOWN

2. a mechanism of fertilization in angiosperms, in which two sperm cells unite with two cells in the embryo sac to form the zygote and endosperm
4. a nutrient-rich tissue that provides nourishment to a developing embryo in angiosperm seeds
6. a structure that develops in the plant ovary and contains the female gametophyte
10. a short stem with four whorls of modified leaves
15. a life cycle that switches between gametophyte and sporophyte stages
16. a symbiotic relationship in which both participants benefit
17. a moss stage with haploid cells
18. a moss, liverwort, or hornwort
19. a system of tube-shaped cells that branches throughout the plant
25. modified leaves, usually green, that enclose a flower before it opens

27. a haploid cell that divides mitotically to produce the gametophyte without fusing with another cell

28. a mutualistic association between a plant root and fungus

30. cone-bearing plants

31. microscopic pores in leaf surfaces that facilitate gas exchange

34. a plant structure that anchors a plant in soil, absorbs and transports minerals and water, and stores food

35. the tube-shaped, nonliving portion of the vascular system in plants that carries water and minerals from the roots to the rest of the plant

41. a pollen-producing part of a flower that consists of a stalk and an anther

42. a mature ovary of a flower that protects dormant seeds and aids in their dispersal

The Evolution of Animals

Studying Advice

a. Many students imagine biology as a chance to learn more about animals that they have seen, read about, watched on TV, or chased away when trying to enjoy time outdoors! If this sounds familiar, then you will enjoy this chapter, which introduces you to the amazing diversity of animal biology. At the end, you will also learn about human evolution and our place within the animal kingdom.

b. Take a big, deep breath before starting this, the longest chapter in the textbook. Chapter 17 introduces animal diversity with a survey of the major invertebrate and vertebrate groups. There is much to consider. Two pieces of advice:

   1) Do not try to study this chapter a night or two before the exam. There are many names and characteristics to learn. You will need time to read, understand, and process this information. Students who wait too long to get started often confuse details on exams.

   2) Try to break this chapter down into smaller, more manageable "bites." Consider reading the sections one at a time, making note cards or other study tools as you read. Take short 5–10 minute breaks between sections.

c. The two organizing tables below will help you categorize the information in many chapter sections. Fill these tables in as you progress through the chapter.

Student Media

Activities

Overview of Animal Phylogeny
Characteristics of Invertebrates
Characteristics of Chordates
Primate Diversity
Human Evolution

Case Studies in the Process of Science

How Are Insect Species Identified?
How Does Bone Structure Shed Light on the Origin of Birds?

MP3 Tutors

Human Evolution

Videos

Discovery Channel Video Clip: Invertebrates
Hydra Eating *Daphnia* (time-lapse)
Thimble Jellies
Hydra Releasing Sperm
Jelly Swimming
C. Elegans Crawling
C. Elegans Embryo Development (time-lapse)
Nudibranchs
Lobster Mouth Parts
Butterfly Emerging
Echinoderm Tube Feet
Manta Ray
Bat Licking Nectar
Gibbons Brachiating

Organizing Tables

Examine Figure 17.6 and the corresponding text pages to complete the table below comparing invertebrate groups. Several cells are filled in to help you in this task.

TABLE 17.1					
Major Animal Phyla	True Tissues	Type of Body Symmetry	Type of Body Cavity	Type of Digestive Tract	Examples of Animals in This Group
Porifera	Yes or No		None	None	
Cnidaria	Yes or No		None		
Platyhelminthes	Yes or No		None		
Nematoda	Yes or No				
Mollusca	Yes or No				
Annelida	Yes or No				
Arthropoda	Yes or No				
Echinodermata	Yes or No				
Chordata	Yes or No				

Examine Figure 17.32 and the corresponding text pages to complete the table below comparing the vertebrate classes. Some of the cells are already filled in to help you in this task.

TABLE 17.2

Vertebrate Groups	Are Jaws Present?	Is the Skeleton Made of Bone or Cartilage?	Do the Adults Have Gills or Lungs?	Do They Have Legs? (If yes, how many do they walk on?)	Do They Have Amniotic Eggs?	Are They Ectothermic or Endothermic?
Jawless vertebrates		Cartilage		No	No	
Chondrichthyans						
Osteichthyans			Most with gills, a few with lungs			
Amphibians			Most with lungs, a few with gills			
Nonbird reptiles						
Birds						
Mammals						

Content Quiz

Directions: Identify the *one* best answer for the multiple-choice questions. For true/false questions, determine if the statement is true or false. If false, change the underlined word(s) to make the statement true. Finally, add the correct word(s) to the fill-in-the-blank questions to make the statements true.

Biology and Society: You're Going to Treat Me with What?!

1. Maggots are approved for use in modern medicine to treat:
   A. sinus infections and sore throats.
   B. headaches.
   C. malignant, cancerous tumors.
   D. arthritis and other diseases of the joints.
   E. diabetic ulcers, bedsores, and burns.

2. True or False?   Leeches secrete a powerful <u>anticoagulant</u> that prevents the clotting of blood.

3. A person who has had a finger reattached surgically might be surprised that the surgeon is using a(n) _____ to promote recirculation in the reattached appendage.

The Origins of Animal Diversity
WHAT IS AN ANIMAL?

4. Which one of the following sets of descriptions best describes animals?
   A. eukaryotic, unicellular, and heterotrophic organisms that use ingestion
   B. eukaryotic, multicellular, and autotrophic organisms that use egestion
   C. eukaryotic, multicellular, and autotrophic organisms that use ingestion
   D. eukaryotic, multicellular, and heterotrophic organisms that use ingestion
   E. prokaryotic, multicellular, and heterotrophic organisms that use ingestion

5. Examine Figure 17.3 showing the development of a sea star. In the early stages of development, the overall size of the embryo grows very little. Therefore, the average size of the embryonic cells:
   A. decreases.
   B. stays about the same.
   C. increases.

6. True or False?   The life histories of many animals include <u>larval</u> stages.

7. True or False?   Animal development typically progresses from <u>zygote</u> to <u>gastrula</u> to <u>blastula</u>.

8. A change in body form, called _____, remodels a larval stage into an adult form.

9. Most animals have a muscular system and a(n) _____ system that controls it.

EARLY ANIMALS AND THE CAMBRIAN EXPLOSION

10. The Cambrian explosion:
    A. was a period of violent volcanic activity.
    B. resulted from Earth's impact with a large comet or asteroid, which killed off the dinosaurs.
    C. refers to a time when animal diversity increased dramatically.
    D. was a huge internal blast deep within Earth that triggered tremendous earthquakes.
    E. resulted in the loss of most animal species and the origin of most plant species.

11. True or False?   The cause of the Cambrian explosion is <u>unknown</u>.

12. True or False?   Most zoologists now agree that the Cambrian organisms can be regarded as ancient representatives of <u>modern</u> animal phyla.

ANIMAL PHYLOGENY

13. Which one of the following phyla does *not* usually show bilateral symmetry?
    A. phylum Platyhelminthes
    B. phylum Arthropoda
    C. phylum Chordata
    D. phylum Annelida
    E. phylum Cnidaria

14. Which one of the following is *not* a function of a body cavity?
    A. cushions internal organs
    B. functions as a hydrostatic organ
    C. provides an extra space to store ingested food
    D. enables internal organs to move independently of the body surface

15. Which one of the following shows radial symmetry?
    A. an apple
    B. a skateboard
    C. a submarine sandwich
    D. a teapot
    E. a T-shirt

16. Most animals that move actively in their environment show _____ symmetry.

17. A body cavity that is only partially lined by tissues derived from mesoderm is a(n) _____.

Major Invertebrate Phyla
SPONGES

18. In sponges, food and oxygen are distributed and wastes are removed by:
    A. a primitive circulatory system with blood.
    B. a water vascular system.
    C. choanocytes.
    D. amoebocytes.
    E. tentacles.

19. Which one of the following is found in sponges?
    A. true tissues
    B. a coelom
    C. a digestive tract
    D. a skeleton
    E. gonads

20. True or False?   Most sponges live in fresh water.

21. Sponge cells called _____ filter bacteria from water.

CNIDARIANS

22. Which one of the following statements about cnidarians is *false*? Cnidarians:
    A. are carnivores.
    B. have a complete digestive tract with mouth and anus.
    C. occur in medusa and/or polyp form.
    D. have cnidocytes.
    E. are mostly marine.

23. Examine Figure 17.4. Which one of the following animals has a body plan that is most similar to the final stage of evolution indicated on the right side of this figure?
    A. a jelly
    B. a shark
    C. an earthworm
    D. a grasshopper
    E. a sponge

24. True or False?   The stinging cells of cnidarians are called <u>choanocytes</u>.

25. A sea anemone is an example of the _____ body plan.

26. A jelly is an example of the _____ body plan.

FLATWORMS

27. Flatworms are:
    A. parasitic or free-living, bilaterally symmetrical, and unsegmented.
    B. free-living, bilaterally symmetrical, and segmented.
    C. free-living, radially symmetrical, and unsegmented.
    D. parasitic, bilaterally symmetrical, and segmented.
    E. parasitic and free-living, radially symmetrical, and segmented.

28. Which of the following animals have a gastrovascular cavity with a single opening?
    A. sponges and flatworms
    B. roundworms and flatworms
    C. jellies and flatworms
    D. roundworms and annelids
    E. sea anemones and annelids

29. True or False?   Tapeworms <u>do not</u> have a mouth or any digestive tract.

30. True or False?   Tapeworms usually have <u>just one</u> host.

31. Tapeworms can infect humans that eat beef that is _____.

ROUNDWORMS

32. Which of the following are traits found in roundworms but not in flatworms?
    A. a gastrovascular cavity and a coelom
    B. a complete digestive tract and a pseudocoelom
    C. a gastrovascular cavity and bilateral symmetry
    D. a mouth and a true coelom
    E. an endoskeleton and a mouth

33. Which one of the following items is shaped most like a roundworm?
    A. a train with hundreds of railroad cars
    B. a round toothpick, tapered at both ends
    C. a spoon
    D. an apple
    E. a pancake

34. True or False? A complete digestive tract allows food to move through the animal in <u>only one</u> direction.

35. The body cavity of roundworms is a(n) _____.

ANNELIDS

36. Which one of the following groups has segmental appendages that help the animals move and exchange gases?
    A. earthworms
    B. polychaetes
    C. leeches
    D. planarians
    E. all of the above

37. Which one of the following traits is found in annelids, but *not* roundworms?
    A. an anus
    B. a complete digestive tract
    C. a mouth
    D. segmentation
    E. bilateral symmetry

38. True or False? Most leeches are <u>parasites</u>.

39. Earthworms eat _____ and eliminate _____, a mixture of undigested material and mucus.

ARTHROPODS

40. Which one of the following is *not* a general characteristic of arthropods?
    A. segmentation
    B. jointed appendages
    C. exoskeleton
    D. six legs
    E. molting

41. Which one of the following habitats, if any, is *not* used by an arthropod?
    A. the surface of other animals
    B. the ocean
    C. underground
    D. the air
    E. All of these habitats are used by at least a few arthropods.

42. More than half of all known species of animals are:

    A. nematodes.

    B. insects.

    C. crustaceans.

    D. vertebrates.

    E. cnidarians.

Matching: Match the group on the left to its best description on the right.

_____ 43. insects

_____ 44. crustaceans

_____ 45. arachnids

_____ 46. millipedes and centipedes

A. adults with 6 legs and usually 1–2 pairs of wings

B. usually 4 pair of walking legs, includes spiders

C. multiple pairs of specialized appendages

D. similar segments along length of long body

47. True or False?   Most animal species on Earth are arthropods.

48. True or False?   Insects with incomplete metamorphosis have a larval stage that looks entirely different from the adult stage.

49. Arthropods must molt their _____ to grow.

50. The most dominant arthropods in the oceans are the _____.

51. The branch of biology called _____ is the study of insects.

52. The many specialized body _____ of arthropods provide an efficient division of labor among body regions.

53. Most arthropods belong to the group called _____.

MOLLUSCS

54. The mollusc body consists of a:

    A. segmented tail, a muscular shield, and a mantle.

    B. muscular mass, a segmented foot, and a shield.

    C. muscular mass, a segmented foot, and a mantle.

    D. muscular foot, a visceral mass, and a mantle.

    E. muscular foot, a segmented visceral mass, and a shield.

55. True or False?   Most gastropods have a shell divided into two halves hinged together.

56. The _____ are a group of molluscs adapted for speed and agility.

57. Many molluscs use a(n) _____, a strap-like rasping organ, to scrape up food.

58. Most molluscs have a shell, secreted by the _____.

59. Molluscs house most of their internal organs in the _____.

ECHINODERMS

60. Echinoderms are:
   A. unsegmented, marine animals with an endoskeleton and water vascular system.
   B. unsegmented, freshwater animals with an endoskeleton and circulatory system.
   C. segmented, freshwater and marine animals with an exoskeleton and circulatory system.
   D. segmented, freshwater and marine animals with an endoskeleton and water vascular system.
   E. segmented, marine animals with an exoskeleton and water vascular system.

61. True or False?   Most echinoderms are <u>sessile,</u> or <u>slow moving</u>.

62. Most adult echinoderms have _____ symmetry, but larva typically have _____ symmetry.

Matching: Match the invertebrate group on the left to its best description on the right.

_____ 63. sponges	A. mollusc group with a single, spiral shell
_____ 64. jellies	B. parasitic flatworms without a digestive tract
_____ 65. corals	C. unsegmented; a complete digestive tract; pseudocoelom
_____ 66. planarians	D. the group with the greatest number of species
_____ 67. blood flukes	E. segmented worm group that is mostly marine
_____ 68. tapeworms	F. a cnidarian group that is a polyp body form
_____ 69. roundworms	G. segmented worm group including bloodsuckers
_____ 70. gastropods	H. mollusc group that includes squids and octopuses
_____ 71. bivalves	I. segmented worm group that increases the fertility of soil
_____ 72. cephalopods	J. a cnidarian group that is a medusa body form
_____ 73. leeches	K. mollusc group with two shells hinged together
_____ 74. earthworms	L. no true tissues; only freshwater or marine habitats
_____ 75. polychaetes	M. free-living flatworm group
_____ 76. arthropods	N. parasitic flatworms that live inside blood vessels

The Vertebrate Genealogy
CHARACTERISTICS OF CHORDATES

77. Which one of the following is *not* a chordate characteristic?
    A. ventral, solid nerve cord
    B. pharyngeal slits
    C. notochord
    D. post-anal tail

78. Examine the evolutionary relationships indicated in Figure 17.6. Which group is most closely related to the phylum Chordata?
    A. molluscs
    B. echinoderms
    C. annelids
    D. cnidarians
    E. arthropods

79. True or False?   The phylum Chordata <u>does not include</u> invertebrates.

80. Other than vertebrates, the phylum Chordata includes _____, which are long and thin and _____, and which as adults are sessile filter feeders.

FISHES

Matching: Match the group on the left to its most distinct feature on the right.

_____ 81. ray-finned fishes
_____ 82. coelacanths
_____ 83. lungfishes
_____ 84. cartilaginous fishes
_____ 85. lampreys

A. have no jaws
B. jaws; swimbladder; deep-sea dwellers thought to be extinct
C. jaws; swimbladder; greatest number of species
D. jaws; gulp air; live in Southern Hemisphere
E. jaws; no swim bladder; skeleton made of cartilage

86. Which one of the following groups was the first to evolve?
    A. chondrichthyans
    B. ray-finned fishes
    C. lobe-finned fishes
    D. jawless vertebrates
    E. amphibians

87. True or False?   If a shark stops swimming it tends to <u>sink</u>.

88. Sharks sense changes in water pressure by using their _____ system.

89. Bony fish have a protective covering over their gills called the _____.

90. When not moving, bony fish do not sink because they have a(n) _____.

AMPHIBIANS

91. Which one of the following characteristics is *not* lost during metamorphosis of a frog?
    A. tail
    B. gills
    C. legs
    D. lateral line system

92. During the life cycle of most frogs, the tadpoles:
    A. and adults are herbivores.
    B. and adults are carnivores.
    C. are herbivores and the adults are carnivores.
    D. are carnivores and the adults are herbivores.

93. Which one of the following groups is most likely to be the direct ancestor of amphibians?
    A. chondrichthyans
    B. ray-finned fishes
    C. reptiles
    D. jawless vertebrates
    E. lobe-finned fishes

94. Animals with four limbs are called <u>hexapods</u>.

95. In addition to lungs, the _____ of most adult amphibians promotes gas exchange.

REPTILES

96. Which of the following were reptilian adaptations for living on land?
    A. amniotic eggs and dry scales
    B. four legs and ears
    C. lungs and a tail
    D. lateral line system and ears
    E. lateral line system and lungs

97. Examine the evolutionary relationships indicated in Figure 17.32. Within the vertebrates, which one of the following is a characteristic common and unique to reptiles, birds, and mammals?
    A. vertebrae
    B. hair
    C. jaws
    D. legs
    E. amniotic eggs

98. True or False?   Because they are ectotherms, reptiles survive on about <u>ten times</u> the calories required by a mammal of similar size.

99. The evolution of the _____ egg allowed reptiles to reproduce without returning to water.

100. Reptiles are _____, absorbing their body heat from the environment.

BIRDS

101. Which one of the following characteristics of birds is *not* found in reptiles?
   A. dry scales
   B. endothermic metabolism
   C. amniotic eggs
   D. lungs

102. Which one of the following is *not* an adaptation for flight in birds?
   A. honeycombed bones
   B. a single ovary
   C. loss of teeth
   D. amniotic eggs
   E. feathers

103. True or False?   The most direct ancestors of birds were small, <u>two-legged dinosaurs</u>.

104. Feathers and the scales of reptiles are made of the protein _____.

MAMMALS

105. Mammals are primarily:
   A. terrestrial and endothermic.
   B. terrestrial and ectothermic.
   C. aquatic and endothermic.
   D. aquatic and ectothermic.

106. Which one of the following is *not* a characteristic of all mammals?
   A. hair
   B. mammary glands
   C. a placenta
   D. endothermy

107. True or False?   A brief gestation followed by development in a pouch is characteristic of the <u>eutherian</u> mammals.

108. The mammals that lay eggs are the _____.

109. Humans, apes, and monkeys belong to the group called _____.

The Human Ancestry
THE EVOLUTION OF PRIMATES

110. Which one of the following is *not* a primate adaptation for living in trees?
    A. rigid shoulder joints
    B. eyes close together in the front of the face
    C. excellent hand-eye coordination
    D. extensive parental care
    E. dexterous hands

111. Which one of the following is a characteristic of Old World but *not* New World monkeys?
    A. ground-dwelling
    B. prehensile tails
    C. single births with a long period of nurturing
    D. nails instead of claws

112. True or False?   In general, apes <u>are larger than</u> monkeys.

113. True or False?   Modern apes live only in tropical regions of the <u>New World</u>.

114. Of all the apes, only gibbons and orangutans are primarily _____.

THE EMERGENCE OF HUMANKIND

Matching: Match the group on the left to its best description on the right.

_____ 115. *Homo habilis*

_____ 116. *Homo neanderthalensis*

_____ 117. *Homo erectus*

_____ 118. *Australopithecus afarensis*

_____ 119. *Homo sapiens*

A. the species of Lucy, about 3.2 million years old

B. the species of modern humans

C. about 2.4 million years old, an early tool user

D. the first species to extend humanity's range beyond Africa

E. large-brained, skilled toolmakers, living 200,000–35,000 years ago

120. *Homo sapiens* appears to have evolved:
    A. about 195,000 years ago in what is now modern Ethiopia.
    B. 500,000 years ago from *Homo neanderthalensis* in what is now modern Europe.
    C. 1.8 million years ago from *Homo erectus* in what is now modern Asia.
    D. 2.4 million years ago from *Australopithecus afarensis* in East Africa.
    E. about 3.2 million years ago in southern Africa.

121. The pattern of the history of human evolution is most like:
    A. a ladder.
    B. a marching band in a parade.
    C. a bush.
    D. people lined up to go through a door.

122. Which one of the following is the correct sequence in the evolution of human culture?
    A. agriculture, hunting and gathering nomads, Industrial Revolution
    B. Industrial Revolution, agriculture, hunting and gathering nomads
    C. hunting and gathering nomads, Industrial Revolution, agriculture
    D. agriculture, Industrial Revolution, hunting and gathering nomads
    E. hunting and gathering nomads, agriculture, Industrial Revolution

123. True or False? Chimps <u>are</u> the parent species of humans.

124. True or False? Most of the distinct humans traits evolved <u>at the same time</u>.

125. True or False? Upright posture evolved in humans <u>before</u> our enlarged brain.

126. True or False? The brains of Neanderthals <u>were slightly larger</u> than brains of modern humans.

127. Fossil evidence indicates that bipedalism evolved at least _____ million years ago.

128. Language, written and spoken, is the main way that _____ is transmitted.

Evolution Connection: Earth's New Crisis

129. Which one of the following human activities is having the greatest impact on other species?
    A. habitat destruction
    B. destruction of the ozone layer
    C. burning fossil fuels
    D. air pollution
    E. erosion

130. True or False? Human cultural evolution is occurring <u>much faster than</u> human biological evolution.

Word Roots

a = without; **gnath** = jaw (agnathans: vertebrates without jaws)

amphi = double; **bio** = life (amphibians: vertebrates that live life in and out of the water)

annel = ring (annelids: segmented worms)

anthrop = man; **oid** = like (anthropoid: monkeys, apes, and humans)

arachn = spider (arachnids: spiders, scorpions, mites, and ticks)

arthro = joint; **pod** = foot (Arthropoda: the phylum of animals with jointed legs)

bi = double; **valva** = leaf of a folding door (bivalves: molluscs with two hinged shells)

centi = one hundred; **ped** = foot (centipedes: arthropods with many legs, one pair per body segment)

ceph(al) = head (cephalopods: molluscs that include octopus and squid)

chondro = cartilage (chondrichthyans: the group of cartilaginous fishes)

echino = spiny; **derm** = skin (Echinodermata: the phylum of sea stars, which have spiny skin)

ecto = outside; **therm** = temperature (ectotherms: animals that absorb heat instead of producing it internally)

endo = inside (endothermic: animals that produce heat internally)

entom = an insect (entomology: the branch of biology that studies insects)

exo = outside (exoskeleton: an external skeleton such as that found in all arthropods)

gaster = belly (gastropods: molluscs with their gut near their muscular foot)

homin = man (hominids: humans and our nearest ancestors)

marsupi = bag or pouch (marsupials: the pouched mammals)

meta = change; **morph** = form (metamorphosis: a change in animal form during a lifetime)

milli = one thousand (millipedes: arthropods with many legs, two pairs per body segment)

mollusc = soft (Mollusca: the phylum of soft-bodied animals usually surrounded by hard shells)

mono = one (monotremes: mammals that lay eggs)

noto = the back; **chord** = string (notochord: a flexible, longitudinal rod characteristic of chordate animals)

osteo = bone; **ichthy** = fish (osteichthyans: bony fishes)

paleo = ancient; **anthrop** = man (paleoanthropology: the study of human evolution)

platy = flat; **helminthes** = a worm (Platyhelminthes: the phylum of flatworms)

por = pore; **fer** = bearer (Porifera: the phylum of sponges)

post = behind (post-anal tail: the type of tail found in chordate animals)

pseudo = false (pseudocoelom: a body cavity that is not completely lined by tissue derived from mesoderm)

tetra = four (tetrapods: the group of terrestrial vertebrates with four legs)

Key Terms

amniotic egg
amphibians
animals
annelids
anthropoids
arachnids
arthropods
bilateral symmetry
birds
bivalves
blastula
body cavity
bony fishes
cartilaginous
  fishes
centipedes
cephalopods
chondrichthyans
chordates
cnidarians
coelom

complete digestive
  tract
crustaceans
culture
dorsal, hollow
  nerve cord
echinoderms
ectotherms
endoskeleton
endotherms
entomology
eutherians
exoskeleton
flatworms
gastropods
gastrovascular
  cavity
hominids
hominoids
ingestion
insects

invertebrates
lancelets
larva
lateral line system
lobe-finned fishes
lungfishes
mammals
mantle
marsupials
medusa
metamorphosis
millipedes
molluscs
molting
monotremes
nematodes
notochord
operculum
osteichthyans
paleoanthropology
pharyngeal slits

placenta
polyp
post-anal tail
primates
pseudocoelom
radial symmetry
radula
ray-finned fishes
reptiles
roundworms
segmentation
sponges
swim bladder
tetrapods
tunicates
vertebrates
water vascular
  system

Crossword Puzzle

Use the Key Terms list from this chapter to fill in the crossword puzzle.

ACROSS

3. the group of chordates that all have a cranium and backbone

5. the type of tail found in chordate animals

6. a body cavity that is not completely lined by tissue derived from mesoderm

8. the study of human evolution

10. the system in a shark that runs along the sides of the body and is sensitive to pressure

12. bony fishes

13. an internal organ in a female mammal that is used to nurture the embryo

14. the most abundant arthropods, they have a three-part body

15. a sexually immature animal

16. lizards, snakes, turtles, and crocodiles

18. the simplest of all animals, they lack true tissues

20. a special type of bony fish with muscular fins supported by stout bones

22. a central digestive compartment with only a single opening

23. the sheet of tissue that secretes the shell of a mollusc

24. the straplike rasping organ of a mollusc

25. short, marine invertebrate chordates with a blade-like shape

27. a change of body form

31. an external skeleton such as that found in all arthropods

33. eukaryotic, multicellular heterotrophs that obtain nutrients by ingestion

35. the simplest animals with bilateral symmetry
36. the type of egg enclosed in a shell and used by reptiles and birds
38. group of bony fish with lungs
40. the sea squirts, invertebrate members of the phylum Chordata
41. segmented worms
42. type of body symmetry identical around a central axis
43. members of the human family
45. snails, oysters, squids, octopuses, and clams
46. animals that produce heat internally
48. arthropods with many legs, one pair per body segment
49. roundworms
52. the group that contains all apes
54. the social transmission of knowledge, customs, beliefs, and arts
55. the group of scorpions, spiders, ticks, and mites
56. animals that absorb heat instead of producing it internally
57. type of body symmetry in which only a single plane cuts the animal into mirror images
58. the floating body plan of a cnidarian
59. the group of molluscs that includes squids and octopuses
60. arthropods with many legs, two pair per body segment
61. feathered endotherms
62. the study of insects
63. a fluid-filled space separating the digestive tract from the outer body wall
64. the type of fishes with a skeleton made of cartilage
65. the type of nerve cord found in the phylum Chordata
66. a body cavity completely lined by tissue derived from mesoderm
67. the group of crabs, lobsters, crayfish, shrimp, and barnacles
68. sea urchins, starfish, and sea cucumbers
69. jellyfish, sea anemones, corals, and hydras

DOWN

1. lorises, pottos, lemurs, tarsiers, and anthropoids
2. the sessile body plan of a cnidarian
4. nematodes
7. the group of terrestrial vertebrates with four legs
9. a flexible, longitudinal rod characteristic of chordate animals
11. animals without backbones
14. eating food
17. a gas-filled sac in bony fish
19. a protective flap that covers the gills of fishes
21. molluscs with a single, spiraled shell
24. the most common type of bony fish
26. the type of system in echinoderms that uses water pressure to move tube feet
28. a member of the vertebrate class that includes frogs and salamanders
29. the division of an animal's body along its length into a series of repeated units
30. the group of pouched mammals, including kangaroos and koalas
32. the group of placental mammals
34. gill openings in the pharynx, characteristic of chordate animals
37. a digestive tube with two openings
39. the primate group that includes monkeys, apes, and humans
44. a group of egg-laying mammals
45. the shedding of an exoskeleton to permit growth
47. an internal skeleton
50. clams, mussels, scallops, and oysters
51. animals with jointed legs and exoskeletons
53. vertebrates with hair
54. cartilaginous fishes
57. an embryonic stage in which the embryo is a hollow ball
63. the type of fishes with a skeleton made of bone
64. animals with a notochord, dorsal hollow nerve cord, pharyngeal slits, and post-anal tail

CHAPTER 18

The Ecology of Organisms and Populations

Studying Advice

Look out a window at some trees and bushes. What determines where any type of plant can and will grow? This chapter begins our consideration of the interaction between living organisms and their environments. It answers some basic questions about why certain plants and animals are restricted to just a few parts of the world and how their numbers are limited. If you enjoy walks through the park or wild areas, this chapter will surely be of interest.

Student Media

Activities

DDT and the Environment
Evolutionary Adaptations
Techniques for Estimating Population Density and Size
Human Population Growth
Analyzing Age-Structure Diagrams
Investigating Survivorship Curves

Case Studies in the Process of Science

Do Pillbugs Prefer Wet or Dry Environments?
How Do Abiotic Factors Affect the Distribution of Organisms?

Graph It

Age Pyramids and Population Growth

Videos

Ducklings
Chimp Cracking Nut

Chimp Agonistic Behavior

Snake Ritual Wrestling

Wolves Agonistic Behavior

Albatross Courtship Ritual

Blue-footed Boobies Courtship Ritual

Giraffe Courtship Ritual

Organizing Tables

Define each of the following levels of ecology using the textbook. These definitions will be a helpful reference as you read and review the chapter.

TABLE 18.1	
Level of Ecology	Textbook Definition
Organismal	
Population	
Community	
Ecosystem	

Compare the three patterns of dispersion in the table below.

TABLE 18.2		
Pattern	Factors That Contribute to This Pattern	Examples
Clumped		
Uniform		
Random		

Content Quiz

Directions: Identify the *one* best answer for the multiple-choice questions. For true/false questions, determine if the statement is true or false. If false, change the underlined word(s) to make the statement true. Finally, add the correct word(s) to the fill-in-the-blank questions to make the statements true.

Biology and Society: The Human Population Explosion

1. The continued increase of the human population in the face of limited resources has led to:
   A. global warming.
   B. civil strife aggravated by depressed economics.
   C. conflicts over oil.
   D. declining health of the oceans.
   E. toxic waste.
   F. all of the above.

2. True or False? The human population on Earth is approaching <u>7 billion</u> people.

3. Earth's most significant biological phenomenon is now the _____.

An Overview of Ecology
ECOLOGY AS SCIENTIFIC STUDY

4. Ecology is the scientific study of interactions between:
   A. organisms and their environment.
   B. different types of animals.
   C. animals and plants.
   D. abiotic factors and the environment.
   E. plants and the organisms that pollinate them.

5. True or False? Many ecologists <u>are able to</u> conduct experiments in the field of ecology.

6. The nonliving chemical and physical factors of the environment are the _____ component.

A HIERARCHY OF INTERACTIONS

Matching: Match the level of ecology on the left to its best description on the right.

_____ 7. organismal ecology

_____ 8. community ecology

_____ 9. ecosystem ecology

_____ 10. population ecology

A. the study of all the interactions between the abiotic factors and the community of species that exists in a certain area

B. the study of the evolutionary adaptations that enable individual organisms to meet the challenges posed by their abiotic environments

C. the study of the factors that affect population size, growth, and composition

D. the study of all the organisms that inhabit a particular area

11. Which one of the following correctly lists the levels of ecology in order of increasingly comprehensive levels?

A. organismal ecology, community ecology, population ecology, ecosystem ecology

B. population ecology, organismal ecology, community ecology, ecosystem ecology

C. ecosystem ecology, community ecology, organismal ecology, population ecology

D. organismal ecology, community ecology, ecosystem ecology, population ecology

E. organismal ecology, population ecology, community ecology, ecosystem ecology

12. True or False?   The distribution of organisms is limited by the abiotic conditions they can tolerate.

13. The _____ is the sum of all the planet's ecosystems, or all of life and where it lives.

ECOLOGY AND ENVIRONMENTALISM

14. The publication of Rachel Carson's book, *Silent Spring* was an important event in the field of ecology because it:

A. detailed the dramatic increases in farm productivity that enabled the United States to grow surplus food and market it overseas.

B. highlighted the worldwide decreases in malaria and other insect-borne diseases due to DDT spraying.

C. focused attention on the genetic resistance to pesticides that evolved in an increasing number of pest populations.

D. detailed the environmental damage caused by the use of DDT.

15. True or False?   The immediate results of spraying DDT were <u>decreases</u> in farm productivity.

16. A basic understanding of the field of _____ is required to analyze environmental issues and plan for better practices.

The Evolutionary Adaptations of Organisms
ABIOTIC FACTORS OF THE BIOSPHERE

17. The patchy nature of the biosphere exists mainly because of:
    A. the types of predatory animals that live in a particular region.
    B. differences in climate and other abiotic factors.
    C. random events due to the chance evolution of a new species in a particular region.
    D. the types of plants that grow in a particular region.

18. Which one of the following is *not* an adaptation to conserve water?
    A. a waxy coating on the leaves and other aerial parts of most plants
    B. a dead layer of outer animal skin containing a water-proofing protein
    C. the ability of human kidneys to excrete a very concentrated urine
    D. All of the above are adaptations to conserve water.

19. Which one of the following is *not* an abiotic factor?
    A. wind
    B. fires, hurricanes, tornadoes, and volcanic eruptions
    C. temperature
    D. predation
    E. water
    F. sunlight

20. True or False?   Most organisms cannot maintain a sufficiently active metabolism at temperatures close to <u>0°C</u>.

21. True or False?   Birds and mammals <u>can</u> remain considerably warmer than their surroundings.

22. True or False?   Most photosynthesis occurs near the <u>bottom</u> of a body of water.

23. True or False?   In some communities, catastrophic disturbances such as fires <u>are necessary</u> to maintain a community.

24. Pollen dispersal, openings in forests, and increased loss of water by evaporation are all consequences of high _____.

25. Adaptations that enable plants and animals to adjust to changes in their environments occur during a period called _____ time.

PHYSIOLOGICAL RESPONSES, ANATOMICAL RESPONSES, BEHAVIORAL RESPONSES

Matching: Match the term on the left to its example on the right.

_____ 26. short-term physiological response

_____ 27. anatomical acclimation

_____ 28. physiological acclimation

_____ 29. behavioral response

A. the gradual growth of a heavy coat of fur on a rabbit as winter approaches

B. producing extra blood cells after a person moves to a home at a higher elevation

C. the sudden appearance of goose bumps as a cold winter wind blasts your skin

D. migrating great distances to reach the best nesting grounds

30. True or False?   Ectotherms are better able to tolerate the greatest temperature extremes.

31. True or False?   In general, animals are more anatomically plastic than plants.

32. Physiological responses that are reversible but take days or weeks to complete are examples of _____.

33. The flagging of the tree in Figure 18.9 is an example of a(n) _____ response.

What Is Population Ecology?

POPULATION DENSITY

Matching: Match the term on the left to its best description on the right.

_____ 34. population ecology

_____ 35. mark-recapture method

_____ 36. population density

_____ 37. population

A. a group of individuals of the same species living in a given area at a given time

B. the field that examines the factors that influence a population's size, density, and characteristics

C. a sampling technique used to estimate wildlife populations

D. the number of individuals of a species per unit area or volume

38. A biologist is trying to determine the number of carp in a pond. She uses a net to capture 100 carp, marks each fish with a tag, and returns them to the pond. A week later she nets out 100 more carp and finds that 50 of them had a tag and were caught the week before. How many carp does she estimate are in the pond?

A. 100

B. 150

C. 200

D. 250

E. 300

39. A biology graduate student uses the mark-recapture method to estimate the population of squirrels living in the middle of the university campus. He sets out 100 traps and finds that each trap has one squirrel inside. He tags the squirrels and releases them. A week later he sets out 100 traps and again catches one squirrel in each trap. In this second set, 75 of the squirrels had tags and were caught last week. When analyzing his data, his research advisor tells him that many students have done this study before and that many of these squirrels seem to be attracted to the traps to get a free meal. The bright graduate student thus concludes that when he makes his calculations using the formula for the mark-recapture method, his estimate will be:

A. pretty accurate, very close to the actual population size.

B. too high compared to the actual population size.

C. too low compared to the actual population size.

40. True or False?   When estimating populations, the <u>larger</u> the number and size of sample plots, the more accurate the estimates of population size.

PATTERNS OF DISPERSION

41. During the springtime, tiger salamanders migrate to small ponds to reproduce. During these mating seasons, these animals show which type of dispersion pattern?

A. clumped

B. scattered

C. uniform

D. random

E. irregular

42. True or False?   Interactions among individuals of a population often produce a <u>random</u> pattern of dispersion.

43. In the absence of strong attractions or repulsions among individuals in a population, we expect to find a(n) _____ pattern of dispersion.

POPULATION GROWTH MODELS

44. Exponential growth typically produces a curve shaped like the letter _____, whereas logistic growth typically produces a curve shaped most like the letter _____.
    A. S . . . J
    B. J . . . S
    C. L . . . S
    D. U . . . J
    E. S . . . L

45. Examine Figure 18.18. From 1945 to 1950, the male fur seal population:
    A. declined.
    B. reached the carrying capacity.
    C. increased.
    D. experienced exponential growth.
    E. crashed.

46. True or False?   Growth rate is greatest in the logistic model at the <u>highest</u> population level.

47. True or False?   Growth rate during exponential growth is greatest at the <u>highest</u> population level.

48. The rate of expansion of a population under ideal conditions is called _____.

49. A description of idealized population growth that is slowed by limiting factors is the _____ model.

50. Environmental factors that restrict population growth are called _____ factors.

REGULATION OF POPULATION GROWTH

51. Which one of the following is most likely a density-dependent factor?
    A. hurricane damage to the nesting sites of birds
    B. a long, very cold winter that kills many bison
    C. a shortage of good browsing vegetation because of overeating
    D. a limited number of nesting sites for birds because of forest fires

52. When deer populations are low,
    A. density-dependent factors have their greatest impact.
    B. many twins are produced.
    C. food quality is poor.
    D. many females fail to reproduce.
    E. density-independent factors have their greatest intensity.

53. Aphids and many other insects often show exponential growth in the spring and then rapid die-offs. These populations are:
    A. crashing when they reach their carrying capacities.
    B. limited by factors closely related to population density.
    C. crashing when they are overcome by disease.
    D. most likely limited by density-independent factors.

54. A population that grows exponentially, but is eventually limited by density-dependent factors, will have a growth chart most similar to the shape of the letter:
    A. J.
    B. U.
    C. L.
    D. S.
    E. V.

55. The boom-and-bust cycles of snowshoe hares are most likely caused by:
    A. competition with other grazing mammals.
    B. fluctuations in the lynx population.
    C. the effects of predation.
    D. fluctuations in the hare's food sources.
    E. the combined effects of predation and fluctuations in the hare's food sources.

56. True or False?   A density-dependent factor intensifies as the population decreases in size.

57. True or False?   Density-independent factors are unrelated to population density.

58. True or False?   In many populations, density-independent factors limit population size before density-dependent factors have their greatest impact.

59. A population that remains near its carrying capacity for a long period of time is most likely limited by _____ factors.

60. The reliance of individuals of the same species on the same limited resources is called _____.

HUMAN POPULATION GROWTH

61. Human population growth:
    A. is closest to the model for exponential growth.
    B. is closest to the model for logistic growth.
    C. has remained steady over the course of human history.
    D. is now on the decline, due to density-dependent factors.
    E. has experienced a pattern of boom and bust.

62. In Italy, slightly negative population growth is occurring because:
    A. individuals younger than reproductive age are relatively underrepresented.
    B. individuals who are beyond reproductive age are in decline.
    C. birth rates remain steady while death rates continue to decline.
    D. birth rates are increasing but death rates are increasing even faster.

63. A unique feature of human population growth is:
    A. the independence of factors affecting the birth rates.
    B. the independence of factors affecting the death rates.
    C. our ability to voluntarily control it.
    D. that it cannot be limited by density-dependent factors.
    E. that it cannot be limited by density-independent factors.

64. True or False?   Since the Industrial Revolution, exponential growth of the human population has resulted mainly from <u>an increase</u> in death rates.

65. True or False?   The population of the United States <u>continues to increase</u>.

66. True or False?   Delayed reproduction dramatically <u>decreases</u> population growth rates.

67. True or False?   Carrying capacity <u>has changed</u> with human cultural evolution.

68. Human population growth is based on the same two general parameters that affect other animal and plant populations: _____ rates and _____ rates.

69. The proportion of individuals in different age groups defines the _____ of that population.

Life Histories and Their Evolution
LIFE TABLES AND SURVIVORSHIP CURVES

70. The survivorship curve for humans has:
    A. the highest mortality later in life.
    B. the highest mortality in the middle of the potential life span.
    C. the highest mortality earliest in life.
    D. an even mortality rate throughout the life span.
    E. high rates of mortality early and late in life and low mortality in the middle of the life span.

71. True or False?   Oysters have a Type III survivorship curve with the <u>highest</u> mortality early in life.

72. Survivorship and mortality in a population can be tracked by a(n) _____.

73. A plot of the number of people still alive at each age in a life table is called a(n) _____.

74. Traits that affect an organism's schedule of reproduction and death make up its _____.

LIFE HISTORY TRAITS AS EVOLUTIONARY ADAPTATIONS

75. Populations that exhibit an equilibrial life history:
    A. have a Type III survivorship curve.
    B. are mostly smaller-bodied species such as insects.
    C. mature later.
    D. produce many offspring.
    E. do not care for their young.

76. True or False?   Life history traits, like anatomical features, <u>are shaped</u> by adaptive evolution.

77. Organisms that exhibit a(n) _____ life history tend to grow exponentially when conditions are favorable.

78. In the locations where guppies were preyed upon by pike-cichlids, which eat mostly large and mature guppies, guppies tend to:
    A. be larger.
    B. mature earlier.
    C. produce fewer offspring each time they give birth.
    D. develop more slowly.

79. True or False?   Reznick and Endler were able to conduct experiments that showed that the main variable affecting the guppy life histories was the <u>presence of different predators</u>.

Evolution Connection: Natural Selection on the American Plains

80. Pronghorn antelopes are able to escape their main predators by:
    A. hiding in deep ravines.
    B. running quickly over great distances.
    C. digging shallow pits in which to hide.
    D. attacking predators with sharp hooves.
    E. leaping great distances over ravines.

81. True or False?   The specific adaptations of pronghorn antelopes <u>may not</u> permit them to live in significantly different environments in other parts of the world.

82. The availability of broadleaf plants and the presence of predators are examples of _____ factors that limit the population of pronghorn antelopes.

Word Roots

a = without; **bio** = life (abiotic component: the nonliving chemical and physical factors in an environment)

bio = life; **sphere** = a ball (biosphere: the sum of all of Earth's ecosystems)

eco = house (ecology: the scientific study of the interactions between organisms and their environments)

intra = within (intraspecific competition: the reliance of individuals of the same species on the same limited resources)

Key Terms

abiotic component
acclimation
age structure
biosphere
biotic component
carrying capacity
clumped
community
community ecology
density-dependent factor

density-independent factor
dispersion pattern
ecology
ecosystem
ecosystem ecology
exponential growth model
growth rate
habitats

intraspecific competition
life history
life table
logistic growth model
mark-recapture method
organismal ecology
population
population density

population ecology
population-limiting factors
random
survivorship curve
uniform

Crossword Puzzle

Use the Key Terms list from this chapter to fill in the crossword puzzle.

ACROSS

6. the study of how members of a population interact with their environment

9. a group of individuals of the same species living in a particular geographic area

11. all the organisms in a given area, along with the nonliving factors with which they interact

12. the type of mathematical model of idealized population growth that is restricted by limiting factors

14. the study of interactions between organisms and their environments

15. the rate of expansion of a population under ideal conditions

16. the type of pattern in which individuals in a population are spaced in a patternless, unpredictable way

17. a type of population-limiting factor whose intensity is unrelated to population size

18. the type of pattern that often results from interactions among individuals of a population

19. the change in population size per time interval

20. the type of ecology concerned with evolutionary adaptations of individual organisms

22. the type of ecology concerned with energy flow and the cycling of chemicals among the various biotic and abiotic factors

23. the global ecosystem

24. the number of individuals in a population that an environment can sustain

26. reversible, long-term physiological responses to the environment

28. all the organisms living together and potentially interacting in a particular area

29. environmental situations in which organisms live

30. a type of population-limiting factor that intensifies as the population increases in size

DOWN

1. the type of competition between individuals of the same species for a limited resource

2. a type of sampling method used to estimate wildlife populations

3. a plot of the number of people still alive at each age

4. the type of component in the environment that is not alive

5. the number of individuals of a species per unit area or volume

7. the traits that effect an organism's schedule of reproduction and death

8. the way individuals are spaced within their area

10. a listing of survival and death in a population in a particular time period

13. a type of environmental factor that restricts population growth

21. a characteristic of a population referring to the proportion of individuals in different age groups

24. the type of ecology concerned with interactions between species

25. the type of component in the environment that is alive

27. describing a dispersion pattern in which individuals are aggregated in patches

CHAPTER 19

Communities and Ecosystems

Studying Advice

Have you ever looked at a pond, the seashore, a forest, or a field and wondered how all of the organisms interact? "Who eats whom?," "Who needs whom?," and "How does the entire system fit together?" are the types of questions addressed by ecologists. Ecosystems are complicated networks full of interactions at so many levels. This chapter, often a favorite of students, explores the many interactions of life in ecosystems.

The many organizing tables below provide a place to arrange the ideas and definitions discussed in this chapter. But, beyond the definitions and details, take time to sit back and think about the complex interactions that keep these systems alive!

Student Media

Activities

Interspecific Interactions

Primary Succession

Energy Flow and Chemical Cycling

Food Webs

Energy Pyramids

The Carbon Cycle

The Nitrogen Cycle

Terrestrial Biomes

Aquatic Biomes

Case Studies in the Process of Science

How Are Impacts on Community Diversity Measured?
How Does Light Affect Primary Productivity?

Graph It

Species Area Effect and Island Biogeography
Animal Food Production Efficiency and Food Policy
Atmospheric CO_2 and Temperature Changes

MP3 Tutors

Trophic Levels and Energy Pyramids

Videos

Coral Reef
Clownfish and Anemone
Discovery Channel Video Clip: Leafcutter Ants
Discovery Channel Video Clip: Rain Forests
Discovery Channel Video Clip: Trees
Discovery Channel Video Clip: Space Plants
Hydrothermal Vent
Tubeworms

Organizing Tables

Describe each of the four properties of a community in Table 19.1 below.

TABLE 19.1	
	Description
Diversity	
Prevalent form of vegetation	
Stability	
Trophic structure	

In Table 19.2, distinguish between the pairs of terms.

TABLE 19.2	
Batesian mimicry vs. Müllerian mimicry	
Parasitism vs. mutualism	
Primary succession vs. secondary succession	

Define and give an example of each of the four trophic levels noted in Table 19.3 below. Try to use examples that could all be found in one food chain.

TABLE 19.3		
	Definition	Example
Producer		
Primary consumer		
Secondary consumer		
Detritivore		

Define and compare the following parts of freshwater biomes in Table 19.4 below, using information from the textbook. Also indicate examples of organisms that live in each zone. One cell in the table is already filled in to help you.

TABLE 19.4

	Definition	Examples of Organisms That Live in This Zone
Freshwater zone		
Photic zone		
Aphotic zone		No plants. Fish and other swimming animals may pass through.
Benthic zone		

In Table 19.5 below, compare the properties of a river or stream near its source versus its properties near the lake or ocean where it empties.

TABLE 19.5

Location in the Stream or River	Water Clarity	Water Temperature	Location of Algae	Types of Benthic Animals	Types of Predators
Near its source					
Where it empties into a lake or the ocean					

Compare the marine biomes in Table 19.6 below.

TABLE 19.6

Marine Biomes	Location	Special Characteristics
Estuaries		
Intertidal zones		
Pelagic zone		
Benthic zone		
Hydrothermal vent communities		
Coral reefs		

Content Quiz

Directions: Identify the *one* best answer for the multiple-choice questions. For true/false questions, determine if the statement is true or false. If false, change the underlined word(s) to make the statement true. Finally, add the correct word(s) to the fill-in-the-blank questions to make the statements true.

Biology and Society: Reefs, Coral and Artificial

1. Which one of the following statements about corals is *false*?
    A. Corals dominate coral reef systems.
    B. Corals secrete soft external skeletons made of sodium chloride.
    C. Corals feed on microscopic organisms and particles of organic debris.
    D. Corals obtain organic molecules from the photosynthesis of symbiotic algae that live in their tissues.

2. True or False?   Some coral reefs cover enormous expanses of <u>deep</u> ocean.

3. Corals are also subject to damage from native and introduced _____.

4. Coral reefs are distinctive and complex _____.

Key Properties of Communities
DIVERSITY, PREVALENT FORM OF VEGETATION, STABILITY, TROPHIC STRUCTURE

Matching: Match the property on the left to its best description on the right.

_____ 5. trophic structure

_____ 6. stability

_____ 7. diversity

_____ 8. prevalent form of vegetation

A. the variety of organisms that make up a community

B. the types and structural features of plants

C. the community's ability to resist change and recover

D. the feeding relationships among the members of the community

9. What property of a community determines the passage of energy and nutrients from plants and other photosynthetic organisms to herbivores and then to carnivores?

A. stability

B. trophic structure

C. diversity

D. prevalent form of vegetation

10. True or False?   A forest dominated by cedar and hemlock trees is a <u>highly stable</u> community because these trees can withstand most natural disturbances, including lightning-caused fires.

11. The term _____, as used by ecologists, considers *both* diversity factors: richness and relative abundance.

12. An organism's _____ includes abiotic factors and other individuals in its population and populations of other species living in the same area.

13. An assemblage of species living close enough together for potential interaction is called a(n) _____.

Interspecific Interactions in Communities
COMPETITION BETWEEN SPECIES

14. What will happen if two species in an ecosystem have identical niches?

A. Both species will be driven to local extinction.

B. One species will be driven to local extinction.

C. One species will be driven to local extinction, or one of the species may evolve enough to use a different set of resources.

D. Both species will evolve enough to use a different set of resources.

E. Both species will be driven to local extinction, or both species will evolve enough to use a different set of resources.

15. Which one of the following typically results when population density increases and nears carrying capacity?
    A. Population growth increases.
    B. Birth rates increase.
    C. Mortality increases.
    D. Individuals have access to a larger share of some limiting resource.

16. According to the competitive exclusion principle:
    A. as the number of species in a community increases, interspecific competition decreases.
    B. species will share the community resources to decrease interspecific competition.
    C. when interspecific competition increases, population densities decrease.
    D. if two species compete for the same limiting resources they cannot coexist in the same place.

17. True or False?   The population growth of a species <u>may be limited</u> by the density of competing species.

18. Interactions between species are called _____ interactions.

19. When populations of two or more species in a community rely on similar limiting resources, they may be subject to _____.

20. An organism's ecological role is the same as its ecological _____.

21. The differentiation of niches that enables similar species to coexist in a community is called _____.

PREDATION

22. Which one of the following statements about predation, according to this chapter, is *false*?
    A. In a predator-prey relationship, the consumer is the predator.
    B. In a predator-prey relationship, the food species is the prey.
    C. In herbivory, the prey is an animal.
    D. Predation is a form of interspecific competition.
    E. Natural selection refines the adaptations of predators and prey.

23. Which one of the following is characteristic of predators?
    A. acute senses
    B. thorns
    C. poisons such as strychnine, mescaline, and tannins
    D. alarm calls
    E. mobbing behavior

24. Which one of the following defense mechanisms would be used by a mother bird to draw attention to herself to keep a predator from attacking her chicks?
    A. alarm calls
    B. active self-defense
    C. fleeing
    D. distraction displays
    E. mobbing

25. If a harmless species is protected by appearing to be a harmful model, it is an example of:
    A. warning coloration.
    B. cryptic coloration.
    C. Müllerian mimicry.
    D. Batesian mimicry.

26. A moth that has a color pattern that makes it appear to blend into its environment is showing:
    A. warning coloration.
    B. cryptic coloration.
    C. Müllerian mimicry.
    D. Batesian mimicry.

27. True or False? Predators that <u>ambush</u> their prey are generally fast and agile.

28. True or False? Many animals use <u>warning</u> coloration to make it difficult to spot them in their environment.

29. True or False? Predator-prey relationships <u>can preserve</u> species diversity.

30. True or False? In <u>Batesian</u> mimicry, two or more unpalatable species resemble each other.

31. A toxic animal will often have a _____ coloration to caution predators.

32. A species that reduces the density of the strongest competitors in a community is called a(n) _____ predator.

SYMBIOTIC RELATIONSHIPS, THE COMPLEXITY OF COMMUNITY NETWORKS

33. Which one of the following is *not* an example of a parasitic relationship?
    A. mosquitoes and people
    B. aphids and plants
    C. root-fungus associations called mycorrhizae
    D. a leech and a fish
    E. tapeworms and cattle

34. Which one of the following is most like a parasitic relationship?
    A. a person shopping for groceries
    B. a student writing a paper for a course
    C. a sailor using wind to propel a sailboat
    D. a student stealing a book from another student
    E. a physician treating the wound of a patient

35. Which one of the following is most like a mutualistic relationship?
    A. a person collecting and eating nuts that have fallen from the trees
    B. giving someone $20 to help you change a flat tire
    C. a person moving from one apartment to another
    D. a cow eating grass
    E. bird migrations over great distances

36. True or False?   In the symbiotic relationship termed mutualism, one organism benefits at the expense of the other.

37. True or False?   Many mutualistic relationships may have evolved from predator-prey or host-parasite interactions.

38. True or False?   Biologists have sorted out the complex networks of many biological communities.

39. In humans and other vertebrates, an elaborate _____ system helps defend the body against parasites.

Disturbance of Communities
ECOLOGICAL SUCCESSION

40. Which one of the following is the most common sequence of appearance of organisms during primary succession?
    A. lichens and mosses, grasses, autotrophic microorganisms, shrubs, and trees
    B. lichens and mosses, grasses, shrubs, and trees, autotrophic microorganisms
    C. autotrophic microorganisms, lichens and mosses, grasses, shrubs, and trees
    D. autotrophic microorganisms, grasses, shrubs, and trees, lichens and mosses

41. Which of the following might happen to a community to cause secondary succession? A disturbance has:
    A. destroyed most of the shrubs and trees but left most microorganisms, lichens, and grasses intact.
    B. destroyed an existing community but left the soil intact.
    C. completely eliminated one level of a community and the soil.
    D. completely destroyed a community and all the components of its environment.
    E. eliminated the abiotic components of a community.

42. True or False?   Ecological disturbances are <u>more common</u> than stability in most communities.

43. The process of community change is called _____.

A DYNAMIC VIEW OF COMMUNITY STRUCTURE

44. Species diversity appears to be greatest in communities with _____ amounts of disturbance.
    A. virtually no
    B. very small
    C. intermediate
    D. very large
    E. continual

45. True or False?   Most communities experience a regular change in <u>species diversity</u>.

46. According to the _____ hypothesis, species diversity appears to be greatest in communities with intermediate degrees of disturbance.

AN OVERVIEW OF ECOSYSTEM DYNAMICS

47. The two key processes of ecosystem dynamics are:
    A. energy flow and chemical recycling.
    B. energy flow and phase changes.
    C. photosynthesis and metabolism.
    D. phase changes and chemical recycling.

48. True or False?   Energy reaches most ecosystems in the form of <u>heat</u>.

49. True or False?   Unlike matter, <u>energy</u> cannot be recycled.

50. The highest level of biological organization is a(n) _____.

51. Chemical elements can be recycled between an ecosystem's living community and the _____ environment.

52. Plants and other producers acquire their carbon, nitrogen, and other chemical elements in inorganic form from the _____ and _____.

53. Energy flow and chemical recycling in an ecosystem depend on the transfer of substances in the _____ structure, or feeding relationships.

TROPHIC LEVELS AND FOOD CHAINS, FOOD WEBS

Matching: Match the term on the left to its best description on the right.

_____ 54. detritivore   A.   an organism that uses photosynthesis

_____ 55. primary consumer   B.   an animal that eats herbivores

_____ 56. producer   C.   decomposer

_____ 57. secondary consumer   D.   herbivore

58. Which of the following *never* function as producers in an ecosystem?
    A. terrestrial plants
    B. fungi
    C. phytoplankton
    D. multicellular algae and aquatic plants

59. All organisms in trophic levels above the producers are:
    A. autotrophic producers.
    B. autotrophic consumers.
    C. heterotrophic producers.
    D. heterotrophic consumers.

60. Examine the food web in Figure 19.23. Which one of the following changes would likely decrease the number of hawks?
    A. an increase in the mouse population
    B. an increase in the snake population
    C. an increase in the owl population
    D. an increase in the lizard population

61. True or False?　The trophic level that supports all others is the <u>consumer</u>.

62. Ecologists divide the species of an ecosystem into different _____ based on their main sources of nutrition.

63. The sequence of food transfer from trophic level to trophic level is called a(n) _____.

64. An ecosystem's main detritivores are _____ and _____.

65. The feeding relationships in an ecosystem are usually woven into elaborate food _____.

Energy Flow in Ecosystems
PRODUCTIVITY AND THE ENERGY BUDGETS OF ECOSYSTEMS, ENERGY PYRAMIDS

66. Which one of the following statements about energy pyramids is true?
    A. Primary consumers form the lowest level of an energy pyramid.
    B. The highest level of an energy pyramid represents the producers.
    C. Most energy pyramids have 10–15 levels.
    D. Herbivores usually appear in the second level of an energy pyramid.

67. Examine the killer whale food chain in Figure 19.21. Within this ecosystem, how would the weight of all the killer whales compare to the collective weight of the zooplankton?
    A. The whale weight would be about 100 times greater.
    B. The whale weight would be about the same as the weight of the zooplankton.
    C. The weight of the zooplankton would be about 10 times greater than the whale weight.
    D. The weight of the zooplankton would be about 100 times greater than the whale weight.
    E. The weight of the zooplankton would be about 1000 times greater than the whale weight.

68. True or False?   Only about 1% of the visible light that reaches producers is converted to chemical energy by photosynthesis.

69. True or False?   On average, only about 10% of the energy in the form of organic matter at each trophic level is stored as biomass in the next level of the food chain.

70. The amount of living organic material in an ecosystem is the _____.

71. The rate at which producers build organic matter is the ecosystem's _____.

ECOSYSTEM ENERGETICS AND HUMAN NUTRITION

72. When we eat beef, we are:
    A. producers.
    B. herbivores.
    C. primary consumers.
    D. secondary consumers.

73. True or False?   It takes about the same amount of photosynthetic productivity to produce <u>10 pounds</u> of corn or one pound of hamburger.

74. In many developing countries, people primarily eat _____ because they cannot afford more energy-expensive foods.

Chemical Cycling in Ecosystems
THE GENERAL SCHEME OF CHEMICAL CYCLING, EXAMPLES OF BIOGEOCHEMICAL CYCLES

75. A chemical's specific route through an ecosystem depends upon the:
    A. particular element and the trophic structure of the ecosystem.
    B. amount of rainfall and the variation of the seasons.
    C. number of producers and consumers.
    D. amount of carbon and nitrogen in the atmosphere.

76. Which one of the following nutrients is not very mobile and is mostly cycled locally?
    A. carbon
    B. nitrogen
    C. phosphorus
    D. water

77. True or False?   Plants get most of their nitrogen from <u>the air</u>.

78. True or False?   Some water can bypass the <u>biotic</u> components of an ecosystem and rely completely on geologic processes.

79. True or False?   Each chemical that moves through an ecosystem cycles through an <u>abiotic</u> <u>reservoir</u>.

80. Chemical cycles in an ecosystem are also called _____ cycles because they include biotic and abiotic components.

Biomes
HOW CLIMATE AFFECTS BIOME DISTRIBUTION, TERRESTRIAL BIOMES

81. The two main factors that determine the type of biome in a particular region are:
    A. temperature and rainfall.

B. wind patterns and light intensity.

C. periodic disturbances and soil quality.

D. predation and proximity to large bodies of water.

82. Most biomes are named:

A. according to the type of wind and type of soil that are predominant in the region.

B. for major physical or climatic features and for their predominant vegetation.

C. after the largest geological feature that is near them, such as a lake, mountain, or ocean.

D. based upon their physical position on Earth.

83. True or False?   If the climates in two geographically separate areas are similar, <u>the same</u> type of biome may occur in them.

84. True or False?   The same type of biome throughout the world <u>has</u> the same species living in it.

85. Species living in the same biome in different parts of the world may look similar to each other because of _____ evolution.

FRESHWATER BIOMES

Matching: Match the term on the left to its best description on the right.

_____ 86. photic zone

_____ 87. aphotic zone

_____ 88. benthic zone

_____ 89. benthos

_____ 90. detritus

A. the substrate at the bottom of all aquatic biomes

B. communities of organisms that live in the benthic zone

C. the area where light levels are too low for photosynthesis

D. shallow water near shore and the upper stratum of water away from shore

E. a major source of food for the benthos

91. Which one of the following types of biomes occupies the largest part of the biosphere?

A. savannas

B. estuaries

C. aquatic biomes

D. grasslands

E. temperate deciduous forest

92. Which one of the following traits is *not* characteristic of a river where it runs into a lake or the ocean?

    A. slow-moving water

    B. warmer water (compared to the water at the source of the river)

    C. clearer water (compared to the water at the source of the river)

    D. predators such as catfish that find food more by scent and taste than by sight

    E. benthic worms and insects that burrow into the muddy bottom

93. True or False? Lakes and ponds that receive large inputs of nitrogen and phosphorus often have <u>low</u> populations of algae.

94. True or False? A stream or river near its source is usually <u>colder</u> and <u>faster</u> than where it reaches a lake or the ocean.

MARINE BIOMES

95. Which one of the following statements about marine biomes is *false*?

    A. Evaporation from the oceans provides most of Earth's rainfall.

    B. Life originated in the sea.

    C. Photosynthesis by zooplankton supplies a substantial portion of the biosphere's oxygen.

    D. Ocean temperatures have a major effect on climate and wind patterns.

    E. Aquatic biomes occupy the largest part of the biosphere.

96. Which one of the following is *not* characteristic of intertidal zones?

    A. twice-daily alternations of submergence in seawater and exposure to air

    B. relatively even temperatures

    C. strong wave action

    D. wide fluctuations in the availability of nutrients

    E. organisms that can burrow or that can cling to rocks or vegetation

97. True or False? Estuaries are crucial feeding areas for <u>many birds</u>.

98. True or False? Many fish and invertebrates use estuaries <u>as breeding grounds</u>.

99. True or False? The major producers in estuaries are <u>saltmarsh grasses and algae</u>.

100. True or False? Marine life is distributed according to <u>different factors than</u> those that affect the distribution of freshwater life.

101. True or False?   The ocean's main photosynthetic producers are zooplankton.

102. The environment where a freshwater stream or river merges with the ocean is called a(n) _____.

103. Fishes, squids, and marine mammals spend most of their time in the _____ zone.

104. The most extensive part of the biosphere is the dark _____ zone, occupying the deepest parts of the ocean.

105. Unusual _____ communities are powered by chemical energy from Earth's interior instead of sunlight.

Evolution Connection: Coevolution in Biological Communities

106. Which one of the following relationships is *not* an example of coevolution?
    A. defenses of plants against herbivores
    B. animal defenses against predators
    C. root-soil relationships
    D. parasite-host relationships
    E. mutual symbiosis

107. The yellow "spots" on the passionflower vine appear to serve two purposes. The yellow "spots":
    A. attract ants and wasps that prey on *Heliconius* eggs or larvae, and the spots discourage *Heliconius* butterflies from laying eggs on leaves.
    B. attract ants and wasps that prey on *Heliconius* eggs or larvae, and the sacs are bags of toxin that kill *Heliconius* butterflies feeding on the leaves.
    C. are sacs of toxin that kill *Heliconius* butterflies feeding on the leaves, and the spots discourage *Heliconius* butterflies from laying eggs on leaves.
    D. are sacs of toxin that kill *Heliconius* butterflies feeding on the leaves, and the spots promote a fungus that kills any *Heliconius* eggs laid on the leaves.
    E. attract ants and wasps that prey on *Heliconius* eggs or larvae, and the spots promote a fungus that kills any *Heliconius* eggs laid on the leaves.

108. True or False?   The evolution of thorns on plants, which limit animal grazing, would be an example of coevolution.

109. The term _____ is used for those cases in which the adaptations of two species are closely connected.

Word Roots

a = without; **photo** = light (aphotic zone: the region of an aquatic ecosystem where light does not penetrate enough for photosynthesis to take place)

benth = the depths of the seas (benthic zone: a seafloor or the bottom of a freshwater, pond, lake, river, or stream)

geo = the Earth (biogeochemical cycles: the various chemical cycles occurring in an ecosystem)

co = together (coevolution: the evolution of two species in response to each other)

crypt = hidden (cryptic coloration: another name for camouflage)

detrit = wear off (detritivores: decomposers in an ecosystem)

eco = house (ecosystem: an ecological system)

herb = grass; **vor** = eat (herbivore: an animal that eats mostly plants and algae)

inter = between (interspecific competition: competition between species)

mutu = reciprocal (mutualism: a symbiosis that benefits both partners)

omni = all (omnivore: an animal that eats plants and animals)

para = near; **site** = food (parasitism: when one organism benefits at the expense of another)

phyto = a plant; **plankto** = drift or wander (phytoplankton: algae and photosynthetic bacteria that drift passively in aquatic environments)

quatr = by fours (quaternary consumer: an organism that eats tertiary consumers)

sym = together; **bio** = life (symbiotic relationship: an interspecific relationship in which one species lives in or on another species)

terti = the third (tertiary consumer: a consumer that eats secondary consumers)

troph = food (trophic structure: the feeding relationships among the various species making up a community)

zo = animal (zooplankton: animals that drift in aquatic environments)

Key Terms

abiotic reservoir
aphotic zone
Batesian mimicry
benthic zone
biogeochemical
  cycles
biomass
biomes
carnivores
chemical cycling
coevolution
community
competitive
  exclusion
  principle
cryptic coloration
decomposers
detritivores
detritus
disturbances

ecological
  succession
ecosystem
energy flow
energy pyramid
estuary
food chain
food webs
herbivores
host
hydrothermal vent
  communities
interspecific
  competition
interspecific
  interactions
intertidal zone
keystone predator
Müllerian mimicry
mutualism

niche
omnivores
parasite
parasitism
pelagic zone
photic zone
phytoplankton
predation
predator
prey
primary
  consumers
primary
  productivity
primary succession
producers
quaternary
  consumers
resource
  partitioning

secondary
  consumers
secondary
  succession
species diversity
species richness
stability
symbiotic
  relationship
temperate zones
tertiary consumers
trophic levels
trophic structure
tropics
warning
  coloration
zooplankton

Crossword Puzzle

Use the Key Terms list from this chapter to fill in the crossword puzzle.

ACROSS

2. the rate at which an ecosystem's plants and other producers build biomass, or organic matter

4. an organism eaten by a predator

6. a type of consumer that eats plants, algae, or autotrophic bacteria

7. a type of consumer that eats secondary consumers

8. a type of ecological succession in which a biological community arises in an area without soil

10. the diverse algae and cyanobacteria that drift passively in the pelagic zone

11. the tendency of a biological community to resist change and return to its original species composition after being disturbed

12. a symbiotic relationship in which the symbiont benefits at the expense of the host

14. a type of ecological succession in which a disturbance has destroyed an existing biological community but left the soil intact

15. a type of consumer that eats tertiary consumers

16. the type of zone that includes the open ocean

20. the type of often bright coloration of animals possessing chemical defenses

21. a type of mimicry that is mutual by two species, both of which are harmful to a predator

23. a type of predator species that reduces the density of the strongest competitors in a community

25. the total number of different species in a community

27. the types of interactions between species

28. the region from the Tropic of Cancer to the Tropic of Capricorn

29. an animal that eats both plants and animals

30. animals that drift in the pelagic zone of an aquatic environment

32. a population's role in its community

36. the reciprocal evolutionary influence between two species

39. all the abiotic factors in addition to the community of species that exist in a certain area

40. a major type of ecosystem that covers a large geographic region

42. a type of chemical cycling occurring in an ecosystem, involving both biotic and abiotic components

45. a type of coloration that is a form of camouflage

46. an animal that eats plants, algae, or autotrophic bacteria

47. nonliving organic matter

50. the nonbiological location where components of biogeochemical cycles are stored

51. the larger participant in a symbiotic relationship, serving as home and feeding ground to the symbiont

52. an animal that eats other animals

53. a force that changes a biological community and usually removes organisms from it

54. organisms that promote the breakdown of organic materials into inorganic ones

55. a type of consumer that eats primary consumers

56. the sequence of food transfer from producers through several levels of consumers in an ecosystem

DOWN

1. the number and relative abundance of species in a biological community

3. an interaction between species in which one species, the predator, eats the other, the prey

5. a type of interspecific relationship in which one species lives in or on another species

9. the type of zone that includes water near shore and at the surface exposed to light

12. an organism that makes organic food molecules from carbon dioxide and water and other inorganic molecules

13. a type of symbiosis that benefits both species

17. a type of principle that states that populations of two species cannot coexist in a community if their niches are nearly identical

18. the feeding relationships in an ecosystem

19. the type of competition between populations of two or more species that require similar limited resources

22. the type of community found on a seafloor and powered by chemical energy

24. latitudes between the tropics and the Arctic Circle or the tropics and the Antarctic Circle

26. the type of zone where land meets the sea

31. the process of biological community change resulting from disturbance

33. the division of environmental resources by coexisting species populations

34. a network of interconnecting food chains

35. all the organisms living together and potentially interacting in a particular area

36. a type of cycling that reuses chemical elements

37. a diagram depicting the cumulative loss of energy from a food chain

38. the amount of organic material in an ecosystem

41. an organism that derives its energy from organic wastes and dead organisms; also called decomposer

42. the type of zone that includes the substrate of a lake, pond, or seafloor

43. the passage of energy through the components of an ecosystem

44. the type of zone which includes water that does not get exposed to light

48. an area where fresh water merges with seawater

49. the type of mimicry in which a species that a predator can eat looks like a species that is harmful

CHAPTER 20

Human Impact on the Environment

Studying Advice

Most of us are interested in conserving our natural environments and willing to make sacrifices to do so. But how shall we concentrate our efforts? Where can we have the greatest impact? This chapter provides a background that helps us find these answers. It may be the most meaningful chapter in the book. As you read, look for the hope and promise of what we can do together.

Student Media

Activities

Fire Ants: An Introduced Species
Water Pollution from Nitrates
DDT and the Environment
The Greenhouse Effect
Madagascar and the Biodiversity Crisis
Conservation Biology Review

Case Studies in the Process of Science

How Are Potential Prairie Restoration Sites Analyzed?

Graph It

Global Fisheries and Overfishing
Municipal Solid Waste Trends in the U.S.
Forestation Change
Global Freshwater Resources
Prospects for Renewable Energy

MP3 Tutors

Global Warming

Videos

Discovery Channel Video Clip: Introduced Species
Discovery Channel Video Clip: Trees

You Decide

Does Human Activity Cause Global Warming?
Can We Prevent Species Extinction?

Organizing Tables

Describe the three main types of biological diversity in Table 20.1 below, noting an example of each.

TABLE 20.1		
	Description	Example
Diversity of ecosystems		
Variety of species		
Genetic variation within each species		

Describe the three main causes of the biodiversity crisis in Table 20.2 below, noting an example of each.

TABLE 20.2		
	Description	Example
Habitat destruction		
Introduced species		
Overexploitation		

Content Quiz

Directions: Identify the *one* best answer for the multiple-choice questions. For true/false questions, determine if the statement is true or false. If false, change the underlined word(s) to make the statement true. Finally, add the correct word(s) to the fill-in-the-blank questions to make the statements true.

Biology and Society: Aquarium Menaces

1. Which one of the following was introduced into the Mediterranean Sea and has already caused severe disruptions of that ecosystem?
   A. the alga *Caulerpa*
   B. morel mushrooms
   C. the bacterium *E. coli*
   D. the northern snakehead fish *Channa argus*
   E. loggerhead sea turtles

2. True or False? Humans have a disproportionately <u>high</u> impact on the environment.

3. Human activities are putting communities, ecosystems, and biological _____ at risk.

Human Impact on Biological Communities
HUMAN DISTURBANCE OF COMMUNITIES

4. Which one of the following statements is *false?*
   A. Much of the United States is now a hodgepodge of early successional growth.
   B. Weedy and shrubby vegetation often colonizes an area after forests are clear-cut and abandoned.
   C. Humans currently use about 30% of Earth's land, mostly for homes and cities.
   D. Most crops are grown in monocultures.
   E. Tropical rain forests are quickly disappearing as a result of clear-cutting for lumber and pastureland.

5. True or False? Human disturbance of communities <u>is limited</u> to the United States.

6. Most crops are grown in _____, intensive cultivations of a single plant variety over large areas.

INTRODUCED SPECIES

7. Which of the following organisms was (were) *accidentally* introduced into a new part of the world?
   A. kudzu
   B. starlings
   C. Australia's rabbits
   D. zebra mussels
   E. corn

8. Which one of the following statements about zebra mussels is *false*? Zebra mussels
   A. were accidentally introduced into the St. Lawrence Seaway in the mid-1980s.
   B. have been the source of viruses that cause respiratory disease in humans.
   C. clog reservoir intake pipes.
   D. compete with native fish species for the plankton used for nourishment.
   E. are now found throughout the Great Lakes and in the Mississippi down to New Orleans.

9. True or False?   Most introduced species <u>fail to thrive</u> in their new homes.

10. True or False?   In many cases, introduced species encounter <u>more</u> pathogens, parasites, and predators than native species.

11. The most common cause of extinction is _____.

12. The second most common cause of extinction is _____.

13. The U.S. Department of Agriculture encouraged the import of _____ to the American South in the 1930s to help control erosion.

Human Impact on Ecosystems

IMPACT ON CHEMICAL CYCLES

Matching: Match the cycles to examples of human activities that disrupt these cycles. Each cycle matches just one choice on the right.

_____ 14. Carbon cycle

_____ 15. Nitrogen and Phosphorus cycles

_____ 16. Water cycle

A. fertilizer and sewage contamination of streams and lakes

B. pumping groundwater for irrigation

C. burning of fossil fuels

17. After completely deforesting one section of the Hubbard Brook Experimental Forest:

    A. net loss of the minerals from the forest remained about the same.

    B. nitrate levels in the drainage creek decreased.

    C. water runoff from the watershed increased by 30–40%.

    D. the root systems of the logged trees still prevented water loss from the soil.

18. Which of the following can cause heavy algal growth if added to water draining into a pond or lake?

    A. carbon

    B. nitrogen compounds

    C. phosphate

    D. nitrogen compounds and phosphate

    E. carbon, nitrogen compounds, and phosphate

19. True or False?   Humans have managed to intrude in one way or another into the dynamics of <u>most</u> ecosystems.

20. Sewage outflow into lakes and rivers can cause "overfertilization" or _____, which can lead to heavy algal growth that suffocates the aerobic life of these ecosystems.

THE RELEASE OF TOXIC CHEMICALS TO ECOSYSTEMS

21. Because of biological magnification, which one of the following members of a food chain will be most affected by the introduction of a toxin into an ecosystem?

    A. producer

    B. herbivore

    C. detritivore

    D. primary consumer

    E. secondary consumer

22. True or False?   Bacteria in the bottom mud of rivers and the ocean convert mercury <u>to an extremely toxic and soluble compound</u>.

23. The toxin _____ was concentrated in herring gull eggs in the Great Lakes food chain, leading to levels nearly 5,000 times higher than that measured in phytoplankton.

24. Which one of the following will likely *not* occur because of global warming?
    A. Plant and animal populations will spread to new locations or die.
    B. Global air temperatures will rise.
    C. Sea levels will decrease as the oceans start to evaporate.
    D. Climate patterns will shift.

25. Which one of the following will likely *not* result from the destruction of the ozone?
    A. an increase in the number of skin cancers
    B. an increase in the number of cataracts
    C. damage to crops and phytoplankton
    D. an increase in global atmospheric temperatures

26. True or False?   Most of the increase in CO_2 levels comes from burning <u>trees</u>.

27. True or False?   If all chlorofluorocarbons were banned today, the chlorine molecules already in the atmosphere <u>would continue</u> to decrease atmospheric ozone levels.

28. The process of global warming because of increased atmospheric levels of carbon dioxide and water vapor in the atmosphere is called the _____.

29. Scientists now routinely find a hole in the ozone layer over the continent of _____.

The Biodiversity Crisis
THE THREE LEVELS OF BIODIVERSITY, THE LOSS OF SPECIES

30. Which one of the following is *not* one of the three main components of biodiversity?
    A. genetic variation within each species
    B. the diversity of ecosystems
    C. the variety of species
    D. the number of individuals of a species

31. Compared to the Cretaceous mass extinction, the current mass extinction is:
    A. broader and faster.
    B. narrower and faster.
    C. narrower and slower.
    D. broader but slower.

32. Which one of the following statements is *false*?

    A. Fewer than 10% of all species have been described by scientists.

    B. About 50% of all known bird species in the world are endangered.

    C. Some researchers estimate that at the current rate of destruction, over half of all plant and animal species will be gone by the end of this new century.

    D. About 20% of the known freshwater fishes in the world either have become extinct during historical times or are seriously threatened.

33. True or False?　The current mass extinction of species is due to the activities of <u>humans</u>.

THE THREE MAIN CAUSES OF THE BIODIVERSITY CRISIS, WHY BIODIVERSITY MATTERS

34. Which one of the following is the single greatest threat to biological diversity?

    A. global warming

    B. depletion of the ozone layer

    C. habitat destruction

    D. introduction of non-native species

    E. overexploitation

35. Pigeons, house sparrows, and starlings are species that have:

    A. gone extinct due to overexploitation.

    B. been impacted by the introduction of the Nile perch.

    C. been introduced by humans.

    D. been severely harmed by habitat destruction.

36. Whales, the American bison, and the Galápagos tortoises are species that have been:

    A. overexploited.

    B. impacted by the introduction of non-native species.

    C. greatly reduced because of the loss of habitat.

    D. poisoned by the introduction of environmental toxins.

37. True or False?　Scientists estimated the average annual value of ecosystem dynamics in the biosphere in the year 1997 at 33 <u>million</u> U.S. dollars.

38. About 25% of all prescription drugs contain substances derived from _____.

39. The introduction of non-native or _____ species often results in the elimination of native species.

Conservation Biology
BIODIVERSITY "HOT SPOTS"

40. The two main goals of conservation biology are to:
    A. preserve individual species and sustain ecosystems.
    B. preserve individual species and create new ecosystems.
    C. identify new species and restore ecosystems.
    D. identify new species and sustain ecosystems.

41. Examine the map of hot spots in Figure 20.17. In general, most hot spots are located:
    A. in the Northern Hemisphere.
    B. in the Southern Hemisphere.
    C. near the poles.
    D. near the equator.

42. True or False?   Biodiversity hot spots are also hot spots of <u>extinction</u>.

43. A species found nowhere else is a(n) _____ species.

CONSERVATION AT THE POPULATION AND SPECIES LEVELS

44. Which one of the following does *not* usually result from habitat fragmentation?
    A. Gene flow increases.
    B. The overall size of populations decreases.
    C. Subpopulations are separated into habitat patches that vary in quality.
    D. The number of sink habitats increases.

45. Conservation biology often highlights the relationships between:
    A. an organism and its environment.
    B. biology and society.
    C. the biotic and abiotic factors of the environment.
    D. biology and Earth.

46. True or False?   The populations of many species were subdivided <u>after</u> humans began altering habitats significantly.

47. True or False?   An area of habitat where a subpopulation's reproductive success exceeds its death rate is called a <u>sink</u> habitat.

48. True or False?   The spotted owl situation illustrates the importance of protecting <u>source</u> habitats.

49. True or False?   <u>Controlled fires</u> were needed to help maintain mature pine trees and to help the red-cockaded woodpecker recover from near-extinction.

50. True or False? <u>Habitat use</u> is typically an issue in conflicts involving conservation biology.

51. A(n) _____ species is in danger of extinction throughout all or a significant portion of its range.

52. A(n) _____ species is likely to become endangered in the foreseeable future throughout all or a significant portion of its geographic range.

CONSERVATION AT THE ECOSYSTEM LEVEL, THE GOAL OF SUSTAINABLE DEVELOPMENT

53. Which one of the following statements about the edges between ecosystems is *false*?
    A. Edges are prominent features of landscapes.
    B. Edges have their own sets of physical conditions.
    C. Edges often have their own type and amount of disturbance.
    D. Edges can have both positive and negative effects on biodiversity.
    E. Landscape edges produced by human activities often have more species.

54. Which one of the following statements about movement corridors is *false*? Movement corridors:
    A. promote dispersal of members of a population.
    B. are often streamside habitats.
    C. are important to species that migrate between different habitats seasonally.
    D. can reduce the spread of disease.

55. True or False? Cowbird populations have <u>decreased</u> due to human activities.

56. True or False? A <u>habitat</u> is a regional assemblage of interacting ecosystems.

57. True or False? The areas surrounding a zoned reserve <u>cannot</u> be used to support the local human population.

58. The application of ecological principles to the study of land-use patterns is called _____.

59. A narrow strip or series of small clumps of quality habitat connecting otherwise isolated patches is a(n) _____.

60. An extensive region of land that includes one or more areas undisturbed by humans is a(n) _____.

61. The goals of _____ are balancing human needs with the health of the biosphere.

Evolution Connection: Biophilia and an Environmental Ethic

62. The concept of _____ reflects a human desire to affiliate with other life in its many forms.

Word Roots

bio = life (biodiversity: all of the variety of life)

eu = good; **troph** = food (eutrophication: overfertilization of a lake)

mono = one (monoculture: intensive cultivation of a single plant variety over a large area)

Key Terms

biodiversity
biodiversity crisis
biodiversity hot
  spot
biological
  magnification
biophilia
conservation
  biology

endangered
  species
endemic species
eutrophication
greenhouse effect
introduced species
landscape

landscape ecology
monocultures
movement
  corridor
ozone layer
population
  fragmentation

sink habitat
source habitat
sustainable
  development
threatened species
zoned reserve

Crossword Puzzle

Use the Key Terms list from this chapter to fill in the crossword puzzle.

ACROSS

4. a small geographic area with an exceptional concentration of species

6. process of atmospheric carbon dioxide trapping heat

7. a human desire to affiliate with other life in its many forms

8. the type of species that is in danger of extinction throughout all or a significant portion of its range

10. the current rapid decline in the variety of life on Earth, largely due to the effects of human culture

11. the accumulation of persistent chemicals in the living tissues of consumers in food chains

12. the type of ecology that examines the application of ecological principles to the study of land-use patterns

13. a species that has a distribution limited to a specific geographic area

16. the type of fragmentation in which populations are split and isolated

17. the band of O_3 in the upper atmosphere that protects life on Earth from the harmful ultraviolet rays in sunlight

18. a type of species likely to become endangered in the foreseeable future throughout all or a significant portion of its geographic range

19. a type of habitat where a species' reproductive success exceeds its death rate and from which new individuals often disperse to other areas

20. a series of small clumps or a narrow strip of quality habitat that connects otherwise isolated patches of quality habitat

21. an extensive region of land that includes one or more areas undisturbed by humans

22. a type of habitat where a species' death rate exceeds its reproductive success

DOWN

1. Earth's great variety of life

2. an increase in productivity of an aquatic ecosystem

3. the type of biology that studies ways to counter the loss of biodiversity

5. the type of species that is moved from its native location to a new geographic region

9. a regional assemblage of interacting ecosystems

14. intensive cultivation of a single plant variety over large areas

15. the type of development that produces long-term prosperity of human societies and the ecosystems that support them

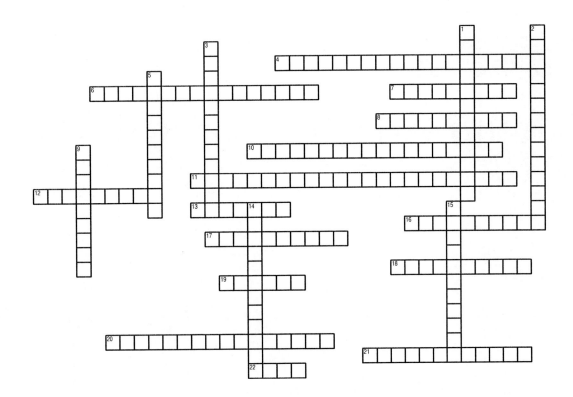

CHAPTER 21

Unifying Concepts of Animal Structure and Function

Studying Advice

a. Before reading, flip through the chapter and look over the figures. This will give you a feel for what the chapter is about and help you identify figures that you will want to refer to again as you read and review.

b. This chapter relates to many activities that occur inside your body. Pause occasionally while reading to reflect on events that you have noticed.

c. Complete the organizing tables below as you read to help define your studies.

Student Media

Activities

Correlating Structure and Function of Cells
The Levels of Life Card Game
Overview of Animal Tissues
Epithelial Tissue
Connective Tissue
Muscle Tissue
Nervous Tissue
Regulation: Negative and Positive Feedback
Human Urinary System
Nephron Function
Control of Water Reabsorption

Case Studies in the Process of Science

How Does Temperature Affect Metabolic Rate in *Daphnia*?
What Affects Urine Production?

Videos

Discovery Channel Video Clip: An Introduction to the Human Body

Organizing Tables

Compare the definitions of the following sets of terms.

TABLE 21.1	
Definition	Definition
Negative feedback:	Positive feedback:
Endotherm:	Ectotherm:
Anatomy:	Physiology:
Open system:	Closed system:
Osmoconformers:	Osmoregulators:
Reabsorption:	Secretion:

Define each of the following types of tissues and provide at least two examples of each.

TABLE 21.2		
Tissue Type	Definition/Key Features	Examples
Epithelia		
Connective		
Muscular		
Nervous		

Describe the main events of the urinary functions performed by the kidneys.

Step	Events
TABLE 21.3	
Filtration	
Reabsorption	
Secretion	
Excretion	

Content Quiz

Directions: Identify the *one* best answer for the multiple-choice questions. For true/false questions, determine if the statement is true or false. If false, change the underlined word(s) to make the statement true. Finally, add the correct word(s) to the fill-in-the-blank questions to make the statements true.

Biology and Society: Keeping Cool

1. Which one of the following will *not* help to cool down the body of an exercising athlete?
   A. lowered blood pressure
   B. widened surface blood vessels
   C. decreased activity
   D. evaporation of sweat

2. True or False? Exercising in a hot environment can lead to intense sweating, dehydration, an eventual drop in blood pressure, and then fainting in a condition called heat <u>stroke</u>.

3. Under extreme conditions of intense heat, a person may suffer from _____ in which the body temperature rises dangerously high, the skin becomes hot and dry, and organs start to fail.

The Structural Organization of Animals
FORM FITS FUNCTION

4. The form/function relationship of biological equipment is a result of:
   A. natural selection.
   B. design.
   C. completely random events.
   D. purposeful planning.

5. True or False? The principle of "form fits function" applies to <u>all levels</u> of biological organization.

6. The study of the <u>structure</u> of an organism and its parts defines _____ and the study of the <u>function</u> of an organism's structural equipment defines _____.

TISSUES

Matching: Match the tissue on the left to its description on the right.

_____ 7. muscular tissue

_____ 8. epithelial tissue

_____ 9. nervous tissue

_____ 10. connective tissue

A. covers the surface of the body and lines organs and cavities within the body, also forms glands

B. found in the brain and spinal cord, cells in this tissue carry signals very rapidly over long distances

C. a sparse population of cells scattered through an extensive extracellular matrix

D. cells have specialized proteins arranged into a structure that contracts when the cell is stimulated

11. Which one of the following is a connective tissue with a matrix that is liquid instead of solid?
    A. adipose tissue
    B. blood
    C. cartilage
    D. bone
    E. fibrous connective tissue

12. Which one of the following is an involuntary muscle tissue found in the walls of the digestive tract and blood vessels?
    A. smooth muscle
    B. cardiac muscle
    C. skeletal muscle
    D. fibrous muscle

13. True or False? The basic unit of nervous tissue is the <u>spinal cord</u>.

14. The most widespread connective tissue in the vertebrate body is _____.

ORGANS AND ORGAN SYSTEMS

15. Which one of the following is *not* a body organ?
    A. brain
    B. blood
    C. stomach
    D. liver
    E. lungs

16. True or False? Two or more tissues packaged into one working unit define an <u>organ</u>.

17. Teams of organs that work together to perform a vital bodily function define _____.

Exchanges with the External Environment

18. Which one of the following statements is *false*?
    A. In an animal body, every living cell must be bathed in a watery solution.
    B. Exchange with the environment is easier for multicellular animals than single-celled organisms.
    C. Simple body forms, such as two-layered sacs and flat shapes, do not allow much complexity in internal organization.
    D. The epithelium of the human lungs has a very large total surface area.
    E. The digestive, respiratory, and urinary systems exchange materials with the external environment.

19. True or False? Every organism is <u>a closed</u> system because all life continuously exchanges chemicals and energy with its surroundings.

20. The cells of more complex animals indirectly exchange materials with the environment by connecting exchange surfaces with body cells through a(n) _____ system.

Regulating the Internal Environment
HOMEOSTASIS

21. Which one of the following is the best example of homeostasis?
    A. the elevation of hormones in the blood during puberty
    B. the elevation of hormones in the blood during pregnancy
    C. an increase in body temperature in response to a bacterial infection
    D. the water content of your cells stays about the same no matter what you drink

22. True or False? All of the body's organ systems play a role in <u>homeostasis</u>.

23. True or False? Homeostasis in the human body permits <u>small</u> changes to occur.

24. The body's tendency to maintain relatively constant conditions in the internal environment despite changes in the external environment defines

_____.

NEGATIVE AND POSITIVE FEEDBACK

25. The part of a negative feedback mechanism in a household furnace that monitors temperature and switches the heater on and off is the:
    A. thermostat.
    B. end zone.
    C. set point.
    D. deflector.
    E. heater.

26. Which of the following represent a positive feedback mechanism?
    A. a healing wound itches, then you scratch it, reinjuring the skin, which itches as it heals
    B. shivering to warm up the body and then stopping after you warm up
    C. eating when you are hungry and stopping when you are full
    D. A & B are examples of a positive feedback mechanism
    E. B & C are examples of a positive feedback mechanism
    F. A, B, & C are examples of a positive feedback mechanism

27. True or False? Negative feedback occurs when the results of some process <u>promote</u> that very process.

28. In _____ feedback, the results of some process inhibit that very process.

THERMOREGULATION

29. Which one of the following represents a behavioral adaptation that assists thermoregulation during cold winter periods?
    A. growing thick fur in the winter
    B. accumulating a thick layer of fat under the skin
    C. increasing the metabolic rate
    D. migrating to suitable climates
    E. None of the above are behavioral adaptations that assist thermoregulation during cold winter periods.

30. Which one of the following does *not* occur when we develop a fever due to a bacterial infection?

    A. bacteria release chemicals called pyrogens
    B. pyrogens travel through the bloodstream to the brain's control center
    C. pyrogens stimulate the control center to raise the body's internal temperature
    D. a mild fever develops, which discourages bacterial growth
    E. All of the above occur when we develop a fever due to a bacterial infection.

31. True or False? Animals that derive the majority of their body heat from their metabolism are called <u>ectotherms</u>.

32. The maintenance of internal body temperature within defined limits defines _____.

OSMOREGULATION

33. Living cells depend on a precise balance of _____ to maintain the integrity of a cell.

    A. salts and minerals
    B. chemical and solar energy
    C. heat and cold
    D. sugars and carbohydrates
    E. water and solutes

34. True or False? Most marine invertebrates are <u>osmoconformers</u>.

35. True or False? All land animals are <u>osmoconformers</u>.

36. The control of the gain or loss of water and dissolved solutes in animals defines _____.

HOMEOSTASIS IN ACTION: THE URINARY SYSTEM

Matching: Match the property on the left to its best description on the right.

_____ 37.	nephron	A.	urine leaves the kidneys in this
_____ 38.	ureter	B.	urine is stored in this
_____ 39.	bladder	C.	the functional unit of the kidneys
_____ 40.	urethra	D.	urine leaves the bladder in this

41. The urinary system plays a central role in homeostasis, forming and excreting
_____ while regulating the amount of _____ in body fluids.
   A. urine, water and salts
   B. blood, sugars
   C. feces, blood
   D. urine, sugars
   E. sweat, water and sugars

42. Which one of the following statements is *false*?
   A. The spleen is the main processing center in the human urinary system.
   B. The urinary system plays a central role in homeostasis.
   C. Every day, the total volume of blood in the body passes into the kidneys
      hundreds of times.
   D. The body uses hormones to control the internal concentration of water
      and dissolved molecules.
   E. Humans with one functioning kidney can lead a normal life.

43. Which of the following is the correct sequence of urinary functions performed
by the kidneys?
   A. reabsorption, filtration, secretion, excretion
   B. secretion, excretion, reabsorption, filtration
   C. secretion, filtration, excretion, reabsorption
   D. filtration, reabsorption, secretion, excretion
   E. filtration, reabsorption, excretion, secretion

44. True or False? The number of kidneys available for transplant is <u>not enough</u> to
meet the current demand for transplants.

45. A person who suffers from kidney failure can be treated by using a blood
filtration machine in a process called _____.

Evolution Connection: How Physical Laws Limit Animal Form

46. The size of single cells is limited by:
   A. physical law.
   B. the size of the other organisms living in the region.
   C. the amount of available sunlight.
   D. the amount of salt in the surrounding water.
   E. the amount of sugar available in the environment.

47. True or False? The independent development of similar forms defines
<u>emergent</u> evolution.

48. The _____ body shape, tapered at both ends, is common in fast-
swimming aquatic animals.

Word Roots

ecto = outside; **therm** = heat (ectotherms: animals that obtain body heat primarily by absorbing it from their surroundings)

endo = inside (endotherms: organisms whose bodies are warmed by heat generated by metabolism; this heat is used to maintain a body temperature higher than that of the external environment)

epi = upon or over (epithelium: tissues that cover the surfaces of the body)

fibro = a fiber (fibrous connective tissue: a dense tissue with large numbers of collagenous fibers organized into parallel bundles)

homeo = same; **stasis** = standing (homeostasis: the steady-state physiological condition of the body)

inter = between (interstitial fluid: the fluid that fills the space between cells)

nephro = the kidney (nephron: the functional unit of a kidney)

neuro = a nerve (neuron: a nerve cell)

Key Terms

adipose tissue
anatomy
bladder
blood
bone
cardiac muscle
cartilage
connective tissue
dialysis
ectotherms
endotherms
epithelial tissue
epithelium

excretion
fever
fibrous connective
  tissue
filtrate
filtration
heat exhaustion
heat stroke
homeostasis
interstitial fluid
loose connective
  tissue
muscle tissue

negative feedback
nephron
nervous tissue
neuron
open system
organ
organ systems
osmoconformers
osmoregulation
osmoregulators
physiology
positive feedback
reabsorption

secretion
skeletal muscle
smooth muscle
thermoregulation
tissue
tubules
ureter
urethra
urine

Crossword Puzzle

Use the Key Terms list from this chapter to fill in the crossword puzzle.

ACROSS

1. the disposal of nitrogen-containing waste products of metabolism

2. the type of dense connective tissue that forms tendons

4. an animal that derives the majority of its body heat from its metabolism

6. an abnormally high internal temperature

9. a tube that releases urine from the body

11. the study of the structure of an organism and its parts

13. a duct leading from the kidney to the urinary bladder

14. type of tissue that stores fat

15. fluid extracted by the excretory system from the blood or body cavity

16. the process by which water and valuable solutes are reclaimed from the filtrate and returned to the blood

17. type of connective tissue consisting of living cells held in a rigid matrix of collagen fibers embedded in calcium salts

18. an animal that has similar water concentrations to its environment

20. category of animal tissue that functions mainly to bind and support other tissues

22. the discharge of specific substances, such as ions and some drugs, into kidney filtrate

25. the type of tissue that is capable of contracting

26. a type of connective tissue with a fluid matrix called plasma

27. the study of the function of an organism's structural equipment

28. a response to overheating characterized by excessive sweating and a quick drop in blood pressure

30. a type of feedback in which a change in some variable triggers mechanisms that amplify the change

31. type of flexible connective tissue found at ends of bones

32. the type of fluid that fills the spaces between cells

33. the most widespread type of connective tissue in the vertebrate body

34. the type of muscle that lacks striations

35. the type of muscle responsible for voluntary movements of the body

36. the type of feedback in which a physiological variable triggers a response that inhibits the same process

37. the steady-state physiological condition of the body

39. the type of system in which blood directly bathes the organs

40. another name for a nerve cell

41. the waste material produced by the vertebrate excretory system

42. the type of tissue made up of neurons and supportive cells

DOWN

1. a type of tissue that forms sheets that cover and line body surfaces

3. an animal that must actively regulate water loss and gain

5. the separation of small blood molecules in which some are kept and others discarded

7. fine tubes in the kidneys

8. the extraction of water and small solutes in the kidney

10. the site of urine storage in the body

12. the maintenance of internal temperature within a narrow range

19. a group of organs that work together to perform vital body functions

20. type of muscle that forms the contractile wall of the heart

21. a specialized center of body function composed of several different types of tissues

23. the control of water balance in organisms

24. a response to overheating characterized by an inability to regulate the body's temperature

29. organisms that obtain body heat primarily by absorbing it from their surroundings

36. the functional unit of the kidney

38. an integrated group of cells with a common structure and function

CHAPTER 22

Nutrition and Digestion

Studying Advice

a. Do you pay attention to the latest news about how to be healthy? Have you ever struggled to lose weight? Do you know someone who has battled an eating disorder? This chapter is full of information related to these and other interesting questions. Some might even say that this chapter is the reward for learning something about chemistry!

b. Use your natural curiosity as motivation to read this chapter and better understand how to live a healthy life.

Student Media

Activities

How Animals Eat Food
Human Digestive System
Analyzing Food Labels
Case Studies of Nutritional Disorders

Case Studies in the Process of Science

What Role Does Salivary Amylase Play in Digestion?

MP3 Tutors

The Human Digestive System

Videos

Shark Eating a Seal
Discovery Channel Video Clip: Nutrition

You Decide

You Decide: Low-fat or Low-carb Diets—Which is Healthier?

Organizing Tables

Compare the definitions and list examples of each of the following types of diet.

TABLE 22.1		
Type of Diet	**Definition**	**Examples of Organisms Using This Type of Diet**
Herbivore		
Carnivore		
Omnivore		

Compare the main events and locations of each of the four stages of food processing in Table 22.2 below.

TABLE 22.2		
Stage	**Description of This Process**	**Digestive Compartment(s) Where This Occurs**
Ingestion		
Digestion		
Absorption		
Elimination		

Compare the different types of essential human nutrients in the table below.

TABLE 22.3		
Type of Nutrient	**Description/Definition**	**Number That Are Essential to Humans**
Essential amino acids		
Vitamins		
Minerals		

Content Quiz

Directions: Identify the *one* best answer for the multiple-choice questions. For true/false questions, determine if the statement is true or false. If false, change the underlined word(s) to make the statement true. Finally, add the correct word(s) to the fill-in-the-blank questions to make the statements true.

Biology and Society: Eating Disorders

1. Which one of the following statements is *false?*
   A. Bulimia involves binge eating followed by purging.
   B. A person with bulimia is usually emaciated like an anorexic person.
   C. The causes of bulimia are unknown.
   D. Bulimia can result in serious health problems.
   E. Bulimia is associated with an obsession with body weight and shape.

2. True or False? We <u>do not</u> know what causes anorexia and bulimia.

3. The condition associated with self-starvation due to an intense fear of gaining weight is called _____.

Overview of Animal Nutrition

Matching: Match the term on the left to its best definition on the right.

_____ 4. elimination   A. eating

_____ 5. absorption   B. the breakdown of food to relatively small nutrient molecules

_____ 6. ingestion

_____ 7. digestion   C. the uptake of small nutrient molecules by the body's cells

   D. the disposal of undigested materials from meals

8. Which one of the following ingests plants and animals?
   A. herbivore
   B. omnivore
   C. carnivore
   D. endovore
   E. maximore

9. Gastrovascular cavities are found in:

   A. cnidarians and flatworms.

   B. reptiles and birds.

   C. many single-celled organisms and some multicellular organisms like sponges.

   D. herbivores but not carnivores.

10. True or False? <u>Mechanical</u> digestion is the chemical breakdown of food by digestive enzymes.

11. In hydrolysis, polymers are broken down into _____.

12. Enzymes that catalyze digestive hydrolysis reactions are called _____.

13. The simplest digestive compartments are _____, intracellular membrane-bound organelles filled with digestive enzymes.

A Tour of the Human Digestive System

Matching: Match each item on the left to its best description on the right.

_____ 14. pepsin

_____ 15. esophagus

_____ 16. *Helicobacter pylori*

_____ 17. gastric juice

_____ 18. small intestine

_____ 19. alimentary canal

_____ 20. duodenum

_____ 21. pancreas

_____ 22. stomach

_____ 23. liver

_____ 24. large intestine

_____ 25. pharynx

_____ 26. microvilli

_____ 27. gallbladder

A. the cause of most stomach ulcers

B. a large sac that can store enough food to sustain us for hours

C. the first section of the small intestine

D. an intersection of the food and breathing pathways

E. where bile is produced

F. a muscular tube that connects the pharynx to the stomach

G. site of water resorption and the formation of feces

H. a digestive fluid secreted by the cells lining the stomach

I. an enzyme that digests proteins

J. another name for the entire digestive tube

K. the longest part of the alimentary canal

L. a gland that secretes pancreatic juice into the duodenum

M. where bile is stored

N. microscopic projections on intestinal cells

28. What keeps the stomach from just eating itself from the inside out?
    A. A protective coating of mucus limits the contact between the cells lining the stomach and the gastric juices.
    B. Nerves and hormones limit the secretion of gastric juice to periods when food is in the stomach.
    C. About once every three days new cells must replace the stomach lining that still gets damaged.
    D. All of the above.
    E. None of the above.

29. Today, most ulcers are effectively cured by:
    A. antacids.
    B. calcium supplements.
    C. antibiotics.
    D. milk.
    E. vitamins.

30. True or False? Occasional backflow of acid chyme into the esophagus causes <u>ulcers</u>.

31. The enzyme salivary _____ hydrolyzes starch.

32. Food is moved through the esophagus and intestines by _____, rhythmic waves of muscular contractions that squeeze the food ball along the esophagus.

33. Rings of muscle that control the movement of materials into and out of the stomach are called _____.

34. The last 6 inches of the large intestine is called the _____, where feces are stored until they can be eliminated.

Human Nutritional Requirements

35. Cellular respiration uses _____ to break down sugar and other food molecules, generating many molecules of _____ for cells to use as a direct source of energy.
    A. oxygen, ATP
    B. carbon dioxide, oxygen
    C. ATP, carbon dioxide
    D. water, oxygen
    E. water, ATP

36. True or False? Dietary calories listed on food labels are actually <u>kilocalories</u>.

37. True or False? Most <u>animal</u> proteins are deficient in one or more of the essential amino acids.

38. True or False? Overdoses of certain <u>vitamins</u> (such as A, D, and K) can be harmful.

39. True or False? Minerals are <u>organic</u> substances required in a healthy diet.

40. The amount of energy it takes just to maintain your basic body support functions is called the _____.

41. Essential amino acids must be obtained in the _____ .

42. Most vitamins function as assistants to _____ in catalyzing metabolic reactions.

43. The minimal standards established by nutritionists for preventing nutrient deficiencies are called the _____.

44. Cells use cellular respiration to extract _____ stored in the organic molecules of food and use it to do work.

45. About 60% of our food energy is lost as _____ that dissipates to the environment.

Nutritional Disorders

46. Which one of the following statements is *false*?
    A. In humans, obesity increases the risk of heart attack, diabetes, and several other diseases.
    B. The best way to maintain a healthy weight is to exercise and eat a balanced diet.
    C. The most reliable sources of essential amino acids are animal products.
    D. Most victims of protein deficiency are adults.
    E. Eating disorders can also cause undernutrition.

47. True or False? <u>Malnutrition</u> is a deficiency in calories.

48. In Africa, the syndrome named kwashiorkor refers to a disease caused by a deficiency of _____.

Evolution Connection: Fat and Sugar Cravings

49. Fat cravings likely evolved in our human ancestors because they were always:
    A. in danger of not finding enough water.
    B. in danger of starvation.
    C. running short of protein.
    D. suffering from obesity.
    E. reproducing.

50. True or False? The majority of Americans consume too many <u>fatty</u> foods, which contribute to obesity.

51. For our ancestors, foods that were _____ or sweet were probably hard to come by.

Word Roots

an = without; **orex** = appetite (anorexia: an eating disorder characterized by self-starvation due to an intense fear of gaining weight)

bulim = hunger (bulimia: a pattern of binge eating followed by purging through induced vomiting)

carni = flesh; **vora** = eat (carnivores: animals that eat other animals)

gastro = stomach; **vascula** = a little vessel (gastrovascular cavities: compartments with a single opening through which food enters and undigested food is expelled)

herb = grass (herbivore: a heterotrophic animal that eats plants)

kilo = a thousand (kilocalorie: a thousand calories)

mal = bad (malnutrition: a deficiency in one or more of the essential nutrients)

omni = all (omnivore: a heterotrophic animal that consumes both meat and plant material)

peri = around; **stalsis** = a constriction (peristalsis: rhythmic waves of contraction of smooth muscle that push food along the digestive tract)

Key Terms

absorption
acid chyme
alimentary canal
anorexia
anus
appendix
basal metabolic
   rate (BMR)
bile
body mass index
   (BMI)
bulimia
calorie
carnivores
chemical digestion
colon

digestion
digestive tubes
duodenum
elimination
esophagus
essential amino
   acids
essential fatty
   acids
essential nutrients
feces
food vacuoles
gallbladder
gastric juice
gastrovascular
   cavities

herbivores
heterotrophs
hydrolases
ingestion
kilocalorie
large intestine
liver
malnutrition
mechanical
   digestion
metabolic rate
minerals
mouth
obesity
omnivores
oral cavity

pancreas
pepsin
peristalsis
pharynx
Recommended
   Daily Allowances
   (RDAs)
rectum
salivary amylase
small intestine
stomach
tongue
vitamins

Crossword Puzzle

Use the Key Terms list from this chapter to fill in the crossword puzzle.

ACROSS

1. the terminal portion of the large intestine where the feces are stored until elimination

4. the final section of the human intestine, it is about 1.5 meters long

5. an organ that stores bile and releases it as needed into the small intestine

8. an enzyme present in gastric juice that begins the hydrolysis of proteins

9. the collection of fluids secreted by the epithelium lining the stomach

11. an area in the vertebrate throat where air and food passages cross

12. a deficiency in one or more of the essential nutrients

14. the simplest type of digestive cavity, found in single-celled organisms

15. another name for the mouth

16. the type of nutrients required for a plant to grow and complete its life cycle

17. the rate of energy consumption of the body

18. a heterotrophic animal that eats plants

20. the initials representing a ratio comparing weight to height

23. the organ that produces bile and detoxifies poisonous chemicals in the blood

25. a gland that secretes digestive enzymes, insulin, and glucagon

26. a tube that connects the pharynx to the stomach

27. a digestive enzyme found in saliva

28. the process of breaking food into molecules small enough for the body to absorb

30. the type of amino acids that an animal cannot synthesize

31. the amount of energy that raises the temperature of 1 gram of water by 1° Celsius

32. an important muscular organ for tasting, chewing, and swallowing

34. enzymes that catalyze digestive hydrolysis reactions

35. the large pouch of the intestine between the esophagus and intestine

36. the uptake of small nutrient molecules by an organism's own body

37. an organic molecule required in the diet in very small amounts

38. a small, fingerlike extension at the junction of the small and large intestine

40. an opening into which food is taken into an animal's body

41. an animal that eats other animals

43. the first section of the small intestine

44. an eating disorder characterized by self-starvation

45. a mixture of substances stored in the gallbladder

46. an eating disorder characterized by binge eating and induced vomiting

47. organisms that must acquire nutrients in the form of organic material from their environment

DOWN

1. abbreviation for the minimal nutritional standards

2. a thousand calories

3. a heterotrophic mode of nutrition in which other organisms or detritus are eaten

6. the type of fatty acids that we cannot produce

7. a heterotrophic animal that consumes meat and plant material

10. the wastes of the digestive tract

13. a type of digestive tract with only a single opening

16. the passing of undigested material out of the digestive compartment

17. in nutrition, a chemical element other than hydrogen, oxygen, or nitrogen that an organism requires for proper body functioning

19. a digestive tract with two openings and one-way flow of food

21. the physical breakdown of food

22. the longest section of the alimentary canal

24. rhythmic waves of contraction that push food along the digestive tract

29. a digestive tract consisting of a tube running between a mouth and an anus

33. the type of digestion characterized by the breakdown of food by enzymes

36. a mixture of recently swallowed food and gastric juice

39. the greatest portion of the large intestine where water is absorbed

42. an inappropriately high body mass index

44. opening through which undigested materials are expelled

45. the abbreviation for the minimal number of kilocalories a resting animal requires to stay alive

CHAPTER 23

Circulation and Respiration

Studying Advice

a. This and the other animal physiology chapters have great relevance to the mechanisms and wonders of your own body. Take the time to notice in yourself the many human reactions discussed in these chapters.

b. As you read, create your own simple flowcharts showing the path of blood through the body and the path of air through the lungs. Then keep these flowcharts handy to clarify events discussed elsewhere in the chapter.

Student Media

Activities

Path of Blood Flow
Cardiovascular System Structure
Cardiovascular System Function
The Human Respiratory System
Transport of Respiratory Gases

Case Studies in the Process of Science

How is Cardiovascular Fitness Measured?

Videos

Discovery Channel Video Clip: Blood

You Decide

Is Second-hand Smoke Dangerous?

Organizing Tables

Compare open and closed circulatory systems in the table below.

TABLE 23.1		
	Open Circulatory System	Closed Circulatory System
The path of blood		
Phyla with this system		

Compare the functions of the four chambers in a human heart in the table below.

TABLE 23.2		
Heart Chamber	Receives Blood From	Sends Blood To
Right atrium		
Right ventricle		
Left atrium		
Left ventricle		

In Table 23.3 below, compare the structures and functions of blood vessels.

TABLE 23.3		
Type of Blood Vessel	Structure of the Walls	Which Ones Have One-Way Valves?
Capillaries		
Arteries		
Veins		

Compare the structures and functions of the components of human blood using the table below.

TABLE 23.4		
Component	Structure	Function(s)
Red blood cells		
White blood cells		
Platelets		

Compare the four main types of respiratory surfaces found in animals using the table below.

TABLE 23.5		
Type of Surface	Structure	Examples of Animals with This System
Skin		
Gills		
Tracheae		
Lungs		

Content Quiz

Directions: Identify the *one* best answer for the multiple-choice questions. For true/false questions, determine if the statement is true or false. If false, change the underlined word(s) to make the statement true. Finally, add the correct word(s) to the fill-in-the-blank questions to make the statements true.

Biology and Society: Doped Up

1. Which one of the following statements about erythropoietin (EPO) of life is *false*? Erythropoietin:
   A. is naturally produced by the body.
   B. is slowly cleared from the bloodstream.
   C. boosts the production of red blood cells.
   D. is sometimes prescribed to relieve the symptoms of anemia.
   E. is abused by athletes to provide an unfair and dangerous advantage.

2. True or False? Erythropoietin increases the production of <u>white</u> blood cells.

3. The circulatory system of most animals works closely with the _____ system.

Unifying Concepts of Animal Circulation
OPEN AND CLOSED CIRCULATORY SYSTEMS

4. Which one of the following is *not* a main component of a circulatory system?
   A. a vascular system
   B. a circulating fluid
   C. lungs or gills
   D. a central pump

5. Which one of the following is the functional center of the circulatory system?
   A. arteries
   B. capillary beds
   C. arterioles
   D. veins
   E. heart

6. True or False? Blood is confined to vessels in <u>a closed</u> circulatory system.

7. True or False? Metabolic wastes, such as <u>oxygen</u>, diffuse from cells into the circulatory system.

8. The closed circulatory system in humans is called a(n) _____ system.

The Human Cardiovascular System
THE PATH OF BLOOD

9. Which one of the following represents the correct sequence of flow of blood through the human cardiovascular system?
   A. left ventricle → pulmonary veins → capillaries in the lungs → pulmonary arteries → right atrium
   B. left ventricle → pulmonary arteries → pulmonary veins → capillaries in the lungs → right atrium
   C. right ventricle → pulmonary veins → capillaries in the lungs → pulmonary arteries → right atrium
   D. right ventricle → pulmonary arteries → capillaries in the lungs → pulmonary veins → left atrium
   E. right atrium → right ventricle → left atrium → capillaries in the lungs → left ventricle

10. True or False? The <u>systemic</u> circuit carries blood from the heart to organs in the rest of the body.

11. Humans and other terrestrial vertebrates have a(n) _____ circulatory system with two distinct circuits of blood flow.

HOW THE HEART WORKS

12. Which one of the following statements about the control of heart rate is *false*?
    A. The SA node generates electrical impulses in the right atrium.
    B. The signals from the SA node can be detected by an EKG.
    C. During a heart attack, the pacemaker is often unable to maintain normal rhythm.
    D. The SA node receives an impulse from the AV node.
    E. The pacemaker of the heart is composed of specialized muscle cells.

13. True or False? Blood in the heart moves only in one direction due to <u>one-way valves</u>.

14. True or False? Caffeine and epinephrine make the heart rate <u>slow down</u>.

15. The relaxation phase of the heart cycle is called _____.

16. A defect in one or more heart valves can lead to a hissing heart sound called a(n) _____.

BLOOD VESSELS

17. Which one of the following statements is *false*?
    A. Veins, but not arteries, have one-way valves.
    B. Veins carry blood that is under very little blood pressure.
    C. Veins have the thinnest walls of any blood vessel.
    D. Arteries have layers of elastic connective tissue and smooth muscle.
    E. Blood pressure is the main force driving the blood from the heart through blood vessels.

18. Blood is primarily moved through veins back to the heart by:
    A. the contraction of surrounding skeletal muscles and one-way valves in veins.
    B. smooth muscle contractions in the walls of capillaries and veins.
    C. the blood pressure created by heart contractions.
    D. the pull of gravity on the blood.
    E. the production of blood throughout the body.

19. True or False? Arteries carry blood <u>away from</u> the heart.

20. True or False? During strenuous exercise, blood is diverted <u>from</u> the digestive tract <u>to</u> skeletal muscles and the skin.

21. Persistent systolic blood pressure above 140 and diastolic pressure above 90 defines _____.

BLOOD

22. Which one of the following is *not* true about red blood cells? Red blood cells:
    A. are the most numerous type of blood cell.
    B. lack nuclei and other organelles.
    C. are well adapted to transport oxygen.
    D. have carbohydrates on their surface that determine the blood type.
    E. contain fibrinogen, an iron-containing molecule that transports oxygen.

23. Which one of the following is *not* true about platelets? Platelets:
    A. are the primary carriers of carbon dioxide in the blood.
    B. adhere to damaged tissue to form a sticky cluster that seals minor breaks.
    C. release clotting factors that convert fibrinogen into fibrin.
    D. help to plug leaks by forming clots.
    E. are bits of cytoplasm pinched off from larger cells in the bone marrow.

24. Compared to red blood cells, white blood cells:
    A. are smaller.
    B. have hemoglobin.
    C. lack nuclei.
    D. are about 1000 times more abundant.
    E. are found outside the circulatory system.

25. True or False? Leukemia is cancer of <u>leukocytes</u>.

26. True or False? An abnormally low amount of hemoglobin or a low number of red blood cells is called <u>anemia</u>.

27. A(n) _____ is a blood clot that forms in the absence of injury.

28. About half of the blood volume is a watery fluid called _____.

29. Most oxygen in blood is bound to _____ inside of red blood cells.

30. Which one of the following is *not* a risk factor for developing cardiovascular disease?
    A. a diet high in cholesterol, trans fat, and saturated fat
    B. a diet high in fruits and vegetables
    C. smoking
    D. lack of exercise

31. True or False? The leading cause of death in the United States is <u>stroke</u>.

32. True or False? The complete blockage of a coronary artery will likely lead to a(n) <u>stroke</u>.

33. The circulatory system contributes to homeostasis by exchanging nutrients and wastes with _____.

34. The chronic cardiovascular disease that results in a narrowing of blood vessels is called _____.

Unifying Concepts of Animal Respiration
THE STRUCTURE AND FUNCTION OF RESPIRATORY SURFACES

35. Which one of the following respiratory systems uses a system of branching tubes to transport oxygen to nearly every cell in the body?
    A. lungs
    B. tracheae
    C. gills
    D. body surface

36. True or False? In general, air holds <u>less</u> oxygen than water.

37. Insects breathe using _____.

38. Extensions, or outfoldings, of the body surface typically used by aquatic organisms are called _____.

39. Cellular respiration uses _____ and glucose to produce energy carrying _____ molecules.

The Human Respiratory System
THE STRUCTURE AND FUNCTION OF THE HUMAN RESPIRATORY SYSTEM

40. As air passes into our respiratory system, it moves from:
    A. pharynx → larynx → trachea → bronchi → bronchioles → alveoli.
    B. pharynx → trachea → larynx → bronchioles → bronchi → alveoli.

C. larynx → pharynx → trachea → bronchioles → bronchi → alveoli.

D. pharynx → larynx → trachea → bronchioles → alveoli → bronchi.

E. trachea → pharynx → larynx → bronchioles → alveoli → bronchi.

41. Which one of the following is *not* considered to be one of the three phases of gas exchange in humans?

A. transport of O_2 from the extensively branched lungs to the rest of the body via the circulatory system

B. production of ATP by aerobic metabolism consuming oxygen and producing carbon dioxide

C. breathing

D. the transport of O_2 from the extensively branched lungs to the rest of the body via the circulatory system

42. True or False? The inner surface of each <u>bronchus</u> is the respiratory surface.

43. Humans' vocal sounds are produced by flexing muscles in the voice box, which stretch the _____ and make them vibrate.

TAKING A BREATH, THE ROLE OF HEMOGLOBIN IN GAS TRANSPORT, HOW SMOKING AFFECTS THE LUNGS

44. Which one of the following is *not* true about blood?

A. Each molecule of hemoglobin consists of four polypeptide chains, each with a heme at the center of which is an atom of iron.

B. Most of the oxygen in blood is carried in hemoglobin molecules.

C. Oxygen readily dissolves in blood.

D. A shortage of iron causes a decrease in the rate of synthesis of hemoglobin.

E. Iron deficiency is the most common cause of anemia.

45. Which one of the following statements about smoking is *false*?

A. One of the worst sources of airborne pollutants is tobacco smoke.

B. The carbon particles in cigarette smoke contains over 4,000 different chemicals.

C. Tobacco smoke kills about 440,000 Americans every year.

D. Smoking causes emphysema.

E. Secondhand cigarette smoke is not considered to be a health hazard.

46. True or False? Hiccups are caused by sudden contractions of the <u>lungs</u>.

47. Expanding the chest cavity to inhale air is known as _____ pressure breathing.

48. Automatic control centers in the brain regulate breathing rate in response to levels of _____ in the blood.

Evolution Connection: The Move onto Land

49. The evolution of vertebrate lungs most likely appeared in:
    - A. reptiles that had gills but could still walk on land.
    - B. aquatic diving birds that also had gills.
    - C. insects that could fly into trees.
    - D. fish that had gills and lungs.
    - E. amphibians already living on land.

50. True or False? Some modern lungfish inhabit stagnant ponds and swamps, which have <u>low</u> levels of oxygen.

51. Comparisons of DNA reveal that the group of vertebrates most closely related to terrestrial vertebrates is the _____.

Word Roots

alveoli = a cavity (alveoli: one of the dead-end, multilobed air sacs that constitute the gas exchange surface of the lungs)

atrio = a vestibule; **ventriculo** = ventricle (atrioventricular node: a region of specialized muscle tissue between the right atrium and right ventricle; it generates electrical impulses that primarily cause the ventricles to contract)

cardi = heart; **vascula** = a little vessel (cardiovascular system: the closed circulatory system characteristic of vertebrates)

erythro = red; **cyte** = cell (erythrocyte: a red blood cell)

fibrino = a fiber; **gen** = produce (fibrinogen: the inactive form of the plasma protein that is converted to the active form fibrin, which aggregates into threads that form the framework of a blood clot)

hyper = above, excessive; **tens** = stretched (hypertension: high blood pressure)

leuk = white; **emia** = blood (leukemia: cancer of leukocytes)

thrombo = a clot (thrombocytes: cells that promote blood clotting)

Key Terms

alveoli
anemia
arteries
arterioles
atherosclerosis
atrium
AV
  (atrioventricular)
  node
blood pressure
breathing
bronchi
bronchioles
capillaries
cardiac cycle
cardiovascular
  disease
cardiovascular
  system

circulatory system
closed circulatory
  system
coronary arteries
diaphragm
diastole
diffusion
double circulation
  system
electrocardiogram
  (ECG or EKG)
embolus
erythrocytes
fibrin
fibrinogen
gills
heart
heart attack
heart murmur

heart rate
hemoglobin
hypertension
larynx
leukemia
leukocytes
lungs
negative pressure
  breathing
open circulatory
  system
pacemaker
pharynx
plasma
platelets
pulmonary circuit
pulse
red blood cells
respiratory surface

respiratory system
SA (sinoatrial)
  node
systemic circuit
systole
thrombocytes
thrombus
trachea
tracheae
veins
ventricle
venules
vocal cords
white blood cells

Crossword Puzzle

Use the Key Terms list from this chapter to fill in the crossword puzzle.

ACROSS

2. the clotting elements in human blood, also called thrombocytes
4. the frequency of heart contractions
6. the type of circuit that transports blood between the heart and the rest of the body
8. a specific type of surface where gases are exchanged between an animal and its environment
10. an iron-containing protein in red blood cells that binds to oxygen
11. the type of node that functions as the pacemaker of the heart
12. another name for white blood cells
13. the typical respiratory surface of animals that live in water
14. the SA node
16. the type of circulatory system in which blood is always contained in blood vessels or the heart
17. the type of disease that affects the heart and or blood vessels
19. the type of system that transports gases and nutrients within the body
20. the contraction stage of the heart cycle, when blood is actively pumped
22. a defect in heart valves that results in a hissing sound
23. a threadlike protein that forms a dense network to create a patch where tissue is damaged
24. the type of arteries that supply the heart
25. the death of heart muscle cells and the resulting failure of the heart
26. the sheet of muscle that expands the chest; it separates the chest cavity from the abdominal cavity
28. the type of blood cells that fight infections
29. the voicebox that contains the vocal cords
30. the plasma protein that is converted into fibrin

32. the force that blood exerts against the walls of blood vessels
35. a muscular organ that pumps blood
37. a heart chamber that receives blood from veins
38. small veins
39. the internal, saclike organ that exchanges gases with the environment
41. a record of the electrical impulses that travel through cardiac muscle during the cardiac cycle
42. another name for platelets
44. alternating processes of inhalation and exhalation
45. in animals, the vessels that return blood to the heart
46. an extensive system of internal tubes that branch throughout an insect's body
48. tiny air-conducting branches that extend from bronchi
50. a type of cardiovascular disease that causes blood vessels to narrow
51. the type of cycle a heart goes through when it contracts rhythmically
53. vessels that carry blood away from the heart
54. the tiniest of all blood vessels
55. the type of circuit that transports blood between the heart and lungs
56. the movement of molecules down a concentration gradient
57. the type of circulatory system that has two distinct circuits of blood flow

DOWN

1. the type of breathing in which air is pulled into the lungs
3. cancer of white blood cells
5. a thrombus that has dislodged from its point of origin and is traveling in blood
7. flexible muscles that vibrate in the larynx, producing vocal sounds

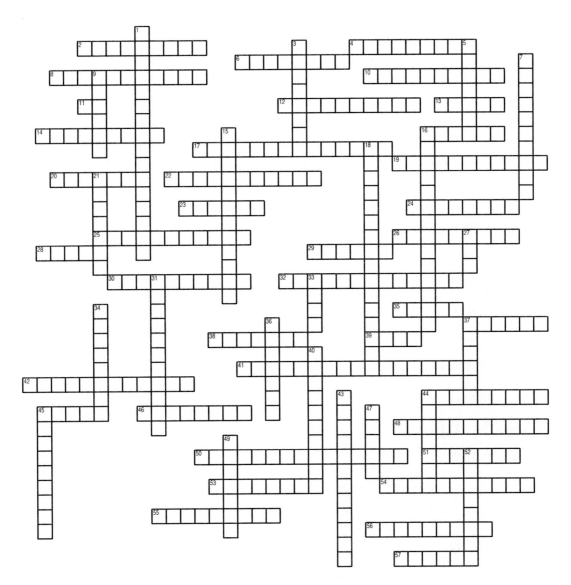

9. the liquid matrix of the blood in which blood cells are suspended

15. another name for red blood cells

16. a type of system with a heart and branching network of blood vessels

18. the type of node that stimulates the ventricles to contract

21. the windpipe

27. the type of blood cells that transport oxygen

31. the type of system that functions in exchanging gases with the environment

33. a type of circulatory system in which blood is pumped through open-ended vessels and out among the body cells

34. another name for a blood clot

36. internal sacs deep within the lungs where gas exchange occurs

37. abnormally low amounts of hemoglobin or a shortage of red blood cells

40. small arteries

43. another name for high blood pressure

44. a pair of breathing tubes that branch off the trachea as they extend to the lungs

45. a heart chamber that pumps blood out of a heart

47. the rhythmic stretching that occurs when blood is forced into arteries

49. the region in the digestive tract that receives food from the oral cavity

52. the stage of the heart cycle in which the heart muscle is relaxed

CHAPTER 24

The Body's Defenses

Studying Advice

Why do we need to get a flu shot every year? Why are some people allergic to certain substances? Why is AIDS so deadly? How does a vaccine prevent disease? Why are we so concerned about a bird flu pandemic? If these questions are ones you have wondered about in the past, this chapter might be of particular interest. The immune system of our body is very complex and involves reactions that few students have ever considered. This chapter is your chance to get a peek inside this hidden and deadly world!

Student Media

Activities

Immune System Responses
HIV Reproductive Cycle

Case Studies in the Process of Science

Why Do AIDS Rates Differ Across the U.S.?
What Causes Infections in AIDS Patients?

MP3 Tutors

The Human Immune System

Videos

T Cell Receptors
Discovery Channel Video Clip: Vaccines

Organizing Tables

Describe the development and functions of the following cells of the immune system.

Cell Type	How It Develops	How It Functions
B cell		
T cell		
Effector cells		
Memory cells		
Helper T cells		
Cytotoxic T cells		

TABLE 24.1

Compare the following pairs of terms, noting similarities and important differences.

External defenses	Internal defenses
Active immunity	Passive immunity
Antigen	Antibody
Humoral immune response	Cell-mediated immune response
Primary immune response	Secondary immune response
Autoimmune diseases	Immunodeficiency diseases

TABLE 24.2

Content Quiz

Directions: Identify the *one* best answer for the multiple-choice questions. For true/false questions, determine if the statement is true or false. If false, change the underlined word(s) to make the statement true. Finally, add the correct word(s) to the fill-in-the-blank questions to make the statements true.

Biology and Society: The Next Pandemic?

1. The rapidly mutating _____ genome of a(n) _____ makes it difficult to develop a vaccine against them, because they evolve so quickly.
   A. DNA, flu virus
   B. DNA, bacterium
   C. RNA, flu virus
   D. RNA, bacterium
   E. Protein, flu virus

2. True or False? In many industrialized nations, childhood vaccination programs have virtually eliminated <u>smallpox and polio</u>.

3. True or False? The flu is responsible for major worldwide outbreaks of disease, called <u>pandemics</u>.

4. One way to prevent infections is to inject a harmless derivative or variant of a disease causing microbe in the form of a _____.

Nonspecific Defenses
EXTERNAL BARRIERS

5. What happens to particles that are trapped in mucous lining the respiratory tract?
   A. Mucuous cells digest the particles.
   B. Acids released by other cells lining the respiratory tract destroy the particles.
   C. The particles are surrounded by a tough protective capsule that prevents infection.
   D. The particles are moved upward until they are swallowed or expelled by coughing, sneezing, etc.
   E. Helper T cells lining the respiratory tract remove the particles.

6. True or False? Concentrated stomach acids <u>kill most of</u> the bacteria swallowed with food or saliva.

7. Lining the mucous membranes of the respiratory and digestive tracts, _____ traps bacteria, dust, and other particles.

8. Sweat, saliva, and tears contain _____, an enzyme that disrupts the cell walls of bacteria.

INTERNAL DEFENSES

9. Which one of the following are white blood cells that penetrate the plasma membranes of virus-infected cells, causing them to burst?
   A. neutrophils
   B. macrophages
   C. natural killer cells
   D. helper T cells
   E. None of the above.

10. True or False? Interferon produced by a virus-infected cell protects other cells from <u>all kinds of viruses</u>, not just of the virus that infected the first cell.

11. Some defensive proteins, called _____ proteins, coat the surfaces of microbes making them easier for macrophages to engulf.

12. Proteins produced by virus-infected body cells that help other cells resist viruses are called _____.

THE INFLAMMATORY RESPONSE

13. Which one of the following causes blood vessels to dilate and leak fluid into the wounded tissue, causing it to swell?
   A. histamine
   B. prostaglandins
   C. interferons
   D. pyrogens
   E. ibuprofen

14. Which one of the following stimulates nerves to send pain signals to the brain?
   A. histamine
   B. prostaglandins
   C. interferons
   D. pyrogens
   E. ibuprofen

15. True or False? Aspirin and ibuprofen treat the symptoms of an illness and <u>address</u> the underlying cause.

16. Chemicals that travel through the bloodstream to the hypothalamus to stimulate a fever are called _____.

THE LYMPHATIC SYSTEM

17. Which one of the following does *not* occur when your body is fighting an infection?
    A. Lymph nodes fill with a huge number of lymphocytes.
    B. Lymph can pick up microbes from infection sites just about anywhere in the body.
    C. Blood delivers microbes to the lymphatic nodes and organs.
    D. Macrophages in lymphatic tissue engulf invaders.
    E. Lymphocytes may be activated to mount a specific immune response.

18. True or False? The lymphatic vessels carry <u>blood</u>, which is similar to interstitial fluid.

19. The lymphatic system includes sac-like organs called _____ that are packed with macrophages and lymphocytes.

Specific Defenses: The Immune System

20. In the United States, widespread vaccination of children has virtually eliminated all of the following except:
    A. smallpox.
    B. polio.
    C. mumps.
    D. measles.
    E. AIDS.

21. In the process of passive immunity, a person becomes resistant to disease by receiving:
    A. antigens.
    B. antibodies.
    C. a live virus.
    D. a harmless variant of a disease-causing microbe or one of its components.
    E. cells that are producing the correct type of antibody for that particular infection.

22. True or False? Antibodies produced against one particular antigen are usually <u>effective</u> against any other foreign substance.

23. True or False? <u>Active</u> immunity is temporary, usually lasting a few weeks or months.

24. A foreign substance that elicits an immune response is a(n) _____.

25. When the immune system detects an antigen, it produces defensive proteins called _____.

26. The term _____ means resistance to specific invaders.

27. In the process of _____, the immune system is confronted with a vaccine composed of a harmless variant of a disease-causing microbe or one of its components.

RECOGNIZING THE INVADERS

28. Which one of the following cell types secretes antibodies?
    A. T cells
    B. B cells
    C. helper T cells
    D. cytotoxic T cells
    E. None of the above.

29. Antibodies are carried to sites of infection in the body by:
    A. humoral secretions.
    B. T cells.
    C. blood.
    D. lymph.
    E. lymph and blood.

30. Which one of the following are involved in cell-mediated immune response *and* humoral immune response?
    A. T cells
    B. B cells
    C. memory cells
    D. effector cells
    E. None of the above.

31. T cells circulate in the blood and lymph, attacking:
    A. body cells that have been infected with bacteria or viruses.
    B. fungi and protozoa.
    C. cancerous cells of our own body.
    D. All of the above.
    E. None of the above.

32. Which one of the following statements about antibodies is *false*?
    A. At the tip of each arm of an antibody, a pair of variable regions forms an antigen-binding site.
    B. Each antibody molecule recognizes and binds to just one specific type of antigen.
    C. An antigen-binding site and the antigen it binds have complementary shapes.
    D. The structural variety of antibodies limits the humoral immune system's ability to react to only a few kinds of antigen.
    E. Antibody binding might cause viruses, bacteria, or foreign eukaryotic cells to form large clumps.

33. True or False? Cell-mediated immune response is produced by <u>effector cells</u>.

34. True or False? Humoral immune response is produced by <u>T cells</u>.

35. Lymphocytes that develop in the bone marrow become _____ cells. Those that develop in the thymus become _____ cells.

36. The overall shape of an antibody is most like the letter _____.

RESPONDING TO THE INVADERS

37. Clonal selection produces a clone of:
    A. helper T cells.
    B. cytotoxic T cells.
    C. memory cells.
    D. effector cells.
    E. None of the above.

38. Which one of the following statements about helper T cells is *false*?
    A. Helper T cells bind to other white blood cells that have previously encountered an antigen.
    B. Helper T cells are activated by the binding of a T cell receptor to a self-nonself complex.
    C. Receptors embedded in a helper T cell's plasma membrane recognize and bind to a complex of self-protein and a foreign antigen displayed on a macrophage.
    D. An activated helper T cell grows and divides, producing more active helper T cells and memory cells.
    E. An activated helper T cell inhibits the activity of cytotoxic T cells.

39. Which one of the following statements about cytotoxic T cells is *false*?
    A. Cytotoxic T cells are the only cells in the body that kill other cells.
    B. Cytotoxic T cells identify infected body cells through binding of a membrane receptor to a self-nonself complex.
    C. An activated cytotoxic T cell discharges the protein perforin that makes holes in the infected cell's plasma membrane.
    D. Another activated cytotoxic T cell protein enters the infected cell and triggers programmed cell death.

40. True or False? The first exposure of lymphocytes to an antigen is called the <u>primary</u> immune response.

41. True or False? Clonal selection produces <u>cytotoxic T</u> cells, which can last decades in the lymph nodes.

42. Effector cells produce _____, specialized for defending against the very antigen that triggered the response.

43. A faster secondary immune response is produced when _____ cells bind an antigen and multiply quickly, producing a large new clone of lymphocytes.

Immune Disorders
ALLERGIES

44. Which one of the following does *not* occur during sensitization?
    A. An allergen enters the bloodstream.
    B. The allergen binds to B cells with complementary receptors.
    C. The B cells then proliferate through clonal selection.
    D. The B cells secrete large amounts of antibodies to that allergen.
    E. The B cells secrete large amounts of histamine that triggers the inflammatory response.

45. True or False? Mast cells release <u>allergens</u>, which causes local inflammation that produces sneezing, coughing, and itching.

46. True or False? Allergic symptoms <u>are most common</u> in the nose and throat because allergens usually enter the body through these passageways.

47. Antigens that cause allergies are called _____.

48. Some people are so extremely sensitive to certain allergens that any contact with them might cause a life-threatening response called _____.

AUTOIMMUNE DISEASES

Match the autoimmune disease on the left to its description on the right.

_____ 49. rheumatoid arthritis

_____ 50. insulin-dependent diabetes

_____ 51. multiple sclerosis

_____ 52. lupus

A. antibodies are produced against histones and DNA

B. damage and painful inflammation of joints

C. destruction of myelin sheaths

D. destruction of cells of the pancreas

53. To prevent organ transplant rejection, doctors
   A. employ cytotoxic T cells to destroy the B cells that will in turn destroy the transplant.
   B. promote anaphylactic shock in the organ recipient.
   C. look for a donor with self-proteins matching the recipient's as closely as possible.
   D. first expose the person about to receive the transplant with antigens found on the organ to be transplanted.
   E. None of the above.

54. True or False? Every individual has a unique set of <u>self-proteins</u>.

55. When the immune system turns against the body's own molecules, _____ diseases occur.

IMMUNODEFICIENCY DISEASES, AIDS

56. Which one of the following statements about AIDS is *false*?
   A. HIV most often attacks helper memory cells.
   B. AIDS is an immunodeficiency disease.
   C. Most people with AIDS die from another infectious agent or from cancer.
   D. Education is the best weapon against the spread of AIDS.
   E. Since 1981, AIDS has killed more than 25 million people worldwide.

57. True or False? People with <u>SCID</u> are born with a marked deficit in T cells and B cells.

58. Hodgkin disease, radiation therapy, and drug treatments used against many cancers can all cause harm by affecting _____, key cells of the immune system.

59. _____ diseases occur in people who lack one or more of the components of the immune system.

60. HIV is deadly because it destroys the _____ system, leaving the body defenseless against most invaders.

61. HIV most often attacks _____ cells, the cells that activate other T cells and B cells.

Evolution Connection: HIV Evolution

62. The greatest obstacle to the eradication of AIDS is
    A. the ability of HIV to hide inside of cells.
    B. the inability to stop its transmission from person to person.
    C. the ability of HIV to trigger anaphylactic shock.
    D. the fast mutation rate of HIV.
    E. its ability to live for long periods outside of the body.

63. True or False? Drug resistant strains of HIV <u>are not</u> found in newly infected patients.

64. HIV now rapidly adapts through the process of _____ in a changing environment.

Word Roots

an = without; **aphy** = suck (anaphylactic shock: an acute, life-threatening allergic response)

anti = against; **gen** = produce (antigen: a foreign macromolecule that does not belong to the host organism and that elicits an immune response)

auto = self (autoimmune diseases: when the immune system turns against the body's own molecules)

cyto = cell; **toxic** = poison (cytotoxic T cells: T cells that kill other cells)

humor = fluid (humoral immune response: the immunity conferred by B cells, which secrete antibodies into blood)

immuno = safe (immune system: the system that protects the body by recognizing and attacking pathogens and cancer cells)

lymph = fluid (lymph: a fluid, similar to interstitial fluid)

macro = large; **phage** = eat (macrophage: an amoeboid cell that moves through tissue fibers, engulfing bacteria and dead cells by phagocytosis)

mono = one (monoclonal antibodies: the antibodies produced by a single clone of cells)

neutro = neutral; **phil** = loving (neutrophils: the most abundant type of leukocyte; neutrophils tend to self-destruct as they destroy foreign invaders, limiting their life span to a few days)

pan = all (pandemic: worldwide outbreaks of disease)

pyro = fire; **gen** = producing (pyrogens: chemicals that stimulate a fever)

Key Terms

active immunity
allergens
allergies
anaphylactic shock
antibody
antigen
autoimmune
 diseases
B cells
cell-mediated
 immune response
clonal selection

complement
 proteins
cytotoxic T cells
effector cells
helper T cells
histamine
humoral immune
 response
immune system
immunity
immunodeficiency
 diseases

inflammatory
 response
interferons
lymph
lymph nodes
lymphatic system
lymphocytes
macrophages
memory cells
monoclonal
 antibodies
natural killer cells

neutrophils
pandemic
passive immunity
primary immune
 response
prostaglandins
pyrogens
secondary immune
 response

Crossword Puzzle

Use the Key Terms list from this chapter to fill in the crossword puzzle.

ACROSS

2. the type of response triggered by penetration of the skin or mucous membranes
6. worldwide outbreaks of disease
9. the system of vessels and lymph nodes, separate from the circulatory system
11. the type of antibodies produced by cells descended from a single cell
13. the type of cells that remain in a lymph node until activated by the antigen that triggered its initial formation
15. a chemical messenger of the immune system that can help other cells resist a virus
16. the type of body system that protects the body by attacking foreign substances
17. organs located along lymph vessels that filter lymph
18. the type of immunity conferred by recovering from an infectious disease
19. molecules that set the body's thermostat to a higher temperature
21. an amoeboid cell that moves through tissue fibers, engulfing bacteria and dead cells by phagocytosis
22. a foreign macromolecule that does not belong to the host organism and that elicits an immune response
25. the type of T lymphocytes that kill infected and cancerous cells
26. the type of nonspecific defensive cells that attack tumor and infected body cells
27. the type of immunity that functions in defense against fungi, protists, bacteria, and viruses inside host cells
29. an antigen that causes allergies
30. one of a group of modified fatty acids secreted by virtually all tissues

31. the type of immunity that uses antibodies to fight bacteria and viruses
33. a set of about 20 serum proteins that carry out a cascade of steps leading to the lysis of microbes
34. an antigen-binding immunoglobulin produced by B cells
35. muscle cells or gland cells that perform the body's responses to stimuli
37. B lymphocytes that are used as a defense mechanism of our immune system

DOWN

1. a type of disease that decreases the body's ability to produce an effective immune response
3. the colorless fluid, derived from interstitial fluid, in the lymphatic system of vertebrate animals
4. the type of temporary immunity obtained by acquiring ready-made antibodies or immune cells
5. the most abundant type of leukocyte
7. the mechanism that determines specificity and accounts for antigen memory in the immune system
8. the type of immune response elicited when an animal encounters an antigen again at some later time
10. resistance to specific invaders through the immune system
12. a white blood cell
14. a substance released by injured cells that causes blood vessels to dilate during an inflammatory response
20. an acute, life-threatening, allergic response
22. a type of immunological disease in which the immune system turns against itself
23. an abnormal sensitivity to an antigen in our surroundings

24. the type of initial immune response to an antigen

28. a harmless variant or derivative of a pathogen used to prevent disease

32. the type of T cell that is required by some B cells to help them make antibodies

36. T lymphocytes that are used as a defense mechanism of our immune system

CHAPTER 25

Hormones

Studying Advice

a. This is a chapter that is very well suited for studying using note cards. The chapter introduces many glands and hormones with specific functions. A note card type of studying system will help you master these fundamentals and organize the information in your mind.

b. As the authors note, Table 25.1 summarizes the human endocrine system. It is a very useful table to refer to often. Mark this page with a piece of paper to make finding it easier as you read.

c. Complete the table below as you read to also help you remember the key functions of each gland. Check the accuracy of what you have written against the information in textbook Table 25.1.

Student Media

Activities

Overview of Cell Signaling

Action of Amino-Acid-Based Hormones

Action of Steroid Hormones

Human Endocrine Glands and Hormones

Case Studies in the Process of Science

How Does a Thyroid Hormone Affect Metabolism?

Videos

Discovery Channel Video Clip: The Endocrine System

Organizing Tables

As you read, complete the table below by describing the functions of each gland.

TABLE 25.1	
Gland	Functions
Hypothalamus	
Anterior pituitary	
Posterior pituitary	
Thyroid	
Parathyroid	
Islet cells of the pancreas	
Adrenal medulla	
Adrenal cortex	
Gonads	

Content Quiz

Directions: Identify the *one* best answer for the multiple-choice questions. For true/false questions, determine if the statement is true or false. If false, change the underlined word(s) to make the statement true. Finally, add the correct word(s) to the fill-in-the-blank questions to make the statements true.

Biology and Society: Of Hunger and Hormones

1. The hormone ghrelin plays a major role in regulating
   A. thirst.
   B. hunger.
   C. pain.
   D. reproductive cycles.
   E. growth.

2. True or False? The regulation and coordination of many systems of the body relies upon _____, chemicals that usually travel in blood and affect cells at distant sites in the body.

3. Hormones help to maintain _____, the steady conditions of the body.

Hormones: An Overview

4. Which one of the following statements about how a steroid hormone works is *false*?
   A. A steroid hormone enters a cell by diffusing through the plasma membrane.
   B. The hormone binds to a receptor protein in the cytoplasm or nucleus.
   C. The hormone triggers a signal-transduction pathway.
   D. The hormone-receptor complex attaches to specific sites on the cell's DNA in the nucleus.
   E. All steroid hormones act by turning genes on or off.

5. Which one of the following is *not* a type of hormone that binds to the plasma membrane of a target cell?
   A. amine hormones
   B. peptide hormones
   C. protein hormones
   D. nucleotide hormones

6. True or False? Hormones are primarily made and secreted by <u>exocrine</u> glands.

7. Derived from cholesterol, _____ hormones bind to receptors inside a cell.

8. A regulatory chemical that travels from its production site through the blood to affect other sites in the body is a(n) _____.

9. Cells that respond to a hormone are _____ cells.

10. The main body system for internal chemical regulation is the _____ system.

The Human Endocrine System

11. Which one of the following has endocrine and nonendocrine functions?
    A. pancreas
    B. pituitary gland
    C. thyroid
    D. cartilage
    E. skeletal muscle

12. True or False? <u>Sex</u> hormones affect most of the tissues of the body.

13. The pituitary gland is controlled by the _____, another endocrine gland that is part of the brain.

THE HYPOTHALAMUS AND PITUITARY GLAND

14. Which one of the following is not released by the pituitary gland?
    A. endorphins
    B. growth hormone
    C. luteinizing hormone
    D. follicle-stimulating hormone
    E. calcitonin

15. Which one of the following statements is *false*?
    A. The anterior pituitary is composed of nonnervous, glandular tissue.
    B. The posterior pituitary is composed of nervous tissue.
    C. The pituitary is the master control center of the entire endocrine system.
    D. Releasing hormones make the anterior pituitary secrete hormones.
    E. Inhibiting hormones make the anterior pituitary stop secreting hormones.

16. True or False? The mammary glands are stimulated to produce milk by <u>endorphins</u>.

17. The hypothalamus makes _____, which signals cells of the kidney to reabsorb water.

18. Serving as the body's painkillers, _____ are made by the anterior pituitary gland.

19. Too little growth hormone during development can lead to _____.

THE THYROID AND PARATHYROID GLANDS

20. Calcium in the body is needed to allow:
    A. muscles to function properly.
    B. nerve signals to be transmitted from cell to cell.
    C. cells to transport molecules across their membranes.
    D. blood to clot.
    E. All of the above.

21. Which of the following requires iodine to produce its hormones?
    A. adrenal medulla
    B. thyroid
    C. parathyroid
    D. pituitary
    E. All of the above.

22. Which of the following might result from excessive secretion of parathyroid hormone?
    A. fatal convulsions known as tetany
    B. gigantism
    C. hypothyroidism
    D. uncontrollable muscle contractions
    E. loss of calcium from bone

23. True or False? Hypothyroidism can result from dietary deficiencies or from a defective thyroid gland.

24. Many industrialized nations have reduced the frequency of _____ by the incorporation of iodine into table salt.

25. Calcitonin and PTH are _____ hormones because they have opposite effects.

26. The hormone _____ causes calcium to be deposited in bones and makes the kidneys reabsorb less calcium as urine is formed.

THE PANCREAS

27. Which of the following best describes the relationship of insulin to glucagon?
    A. They work together to prepare the body to deal with stress.
    B. Insulin stimulates the pancreas to secrete glucagon.
    C. High levels of insulin inhibit pancreatic secretion of glucagon.
    D. They are antagonistic hormones.
    E. Insulin is a steroid hormone; glucagon is a protein hormone.

28. True or False? When the concentration of glucose in the blood rises, following the digestion of a meal, insulin is released.

29. A rise in the concentration of sugar in the blood is caused by the hormone _____.

30. In the disease _____, body cells are unable to absorb glucose from the blood.

31. Type I diabetes requires regular supplements of _____.

32. Treatment for Type II diabetes includes exercise and controlling the intake of _____ in the diet.

THE ADRENAL GLANDS

33. Which one of the following is *not* a response to epinephrine?
    A. release of glucose from the liver
    B. increased heart rate
    C. increased blood pressure
    D. increased absorption of glucose by the digestive tract
    E. increased breathing rate

34. Which one of the following glands is located nearest to the kidneys?
    A. pancreas
    B. pituitary gland
    C. parathyroid glands
    D. testes
    E. adrenal glands

35. True or False? Epinephrine and norepinephrine are secreted by the cells in the adrenal cortex.

36. Receiving _____ for a long period of time makes a person highly susceptible to infection.

THE GONADS

37. Which, if any, of the following is *not* a category of sex hormone?
    A. prolactin
    B. estrogens
    C. androgens
    D. progestins
    E. All of the choices *are* sex hormones.

38. Which, if any, of the following would *not* be secreted in a sexually mature, healthy woman before she reaches menopause?

    A. testosterone

    B. estrogens

    C. progestins

    D. luteinizing hormone

    E. All of the choices *would* be secreted.

39. True or False? Females and males <u>have</u> all three types of sex hormones, but in different proportions.

40. The uterus is prepared to support a developing embryo by the hormones called _____.

41. The female reproductive system is maintained by the hormones called _____.

42. The development and maintenance of the male reproductive system is stimulated by the hormones called _____.

Matching: Match the gland on the left to its best description on the right.

_____ 43.	parathyroid	A. stores and releases antidiuretic hormone
_____ 44.	anterior pituitary	B. releases PTH to help regulate calcium levels in blood
_____ 45.	gonad	
_____ 46.	posterior pituitary	C. produce insulin and glucagon
_____ 47.	adrenal medulla	D. produces glucocorticoids
_____ 48.	thyroid	E. produces hormones that regulate metabolic rate
_____ 49.	islet cells	F. synthesizes and releases FSH, LH, GH, and PRL
_____ 50.	adrenal cortex	G. secretes androgens, estrogens, and progestins
		H. secretes epinephrine and norepinephrine

Evolution Connection: The Changing Roles of Hormones

51. The hormone prolactin is present in many kinds of vertebrates, but it controls different processes in different vertebrate groups. What does this situation say about the probable evolutionary history of this hormone?

    A. Prolactin has probably evolved several times among the vertebrates.

    B. Prolactin is probably evolutionarily ancient and has been turned to different uses in different groups.

    C. Prolactin was probably the "master hormone" of primitive vertebrates.

    D. The prolactin gene is probably highly susceptible to mutation.

    E. The prolactin gene probably resides within mitochondria.

52. Which one of the following is *not* a function of prolactin?

    A. It stimulates mammary glands to grow and produce milk in mammals.

    B. In amphibians, it stimulates movement towards water and affects metamorphosis.

    C. It helps to regulate salt and water balance in fish that migrate between salt and fresh water.

    D. In birds, it regulates calcium and glucose levels in the blood.

53. True or False? Hormones play important roles in <u>virtually all</u> organisms.

Word Roots

andro = male; **gen** = produce (androgens: the principal male steroid hormones, such as testosterone, which stimulate the development and maintenance of the male reproductive system and secondary sex characteristics)

endo = within (endorphins: the body's natural painkillers)

epi = above, over (epinephrine: a hormone produced as a response to stress; also called adrenaline)

gluco = sweet (glucagon: a hormone that increases blood sugar levels)

hypo = below (hypothalamus: the master control center of the endocrine system, located just below the thalamus in the brain)

lut = yellow (luteinizing hormone: a gonadotropin secreted by the anterior pituitary)

mellit = honey (diabetes mellitus: a hormonal disease in which cells cannot obtain glucose from blood, resulting in high glucose levels in blood and urine)

para = beside (parathyroid glands: important endocrine glands located next to the thyroid gland)

pro = before; **lact** = milk (prolactin: a hormone produced by the anterior pituitary gland; it stimulates milk synthesis in mammals)

Key Terms

adrenal cortex	endorphins	hypothalamus	pituitary gland
adrenal glands	epinephrine	insulin	posterior pituitary
adrenal medulla	estrogens	islet cells	progestins
androgens	follicle-	luteinizing	prolactin (PRL)
antagonistic	stimulating	hormone (LH)	steroid hormones
hormones	hormone (FSH)	norepinephrine	target cell
anterior pituitary	glucagon	pancreas	thyroid gland
calcitonin	glucocorticoids	parathyroid glands	
corticosteroids	gonads	parathyroid	
diabetes mellitus	growth hormone	hormone (PTH)	
endocrine glands	(GH)		
endocrine system	hormone		

Crossword Puzzle

Use the Key Terms list from this chapter to fill in the crossword puzzle.

ACROSS

1. the ventral part of the vertebrate forebrain that helps maintain homeostasis
4. a hormone that stimulates milk production in mammals
5. a type of lipid hormone made from cholesterol
6. a family of steroid hormones, including progesterone, produced by the mammalian ovary
10. the endocrine glands that secrete a hormone that raises blood calcium levels
11. the type of cells that respond to a regulatory signal, such as a hormone
13. the male or female sex organ
15. the endocrine gland that secretes iodine-containing hormones
16. a vertebrate hormone that lowers blood glucose levels by promoting the uptake of glucose by most body cells
17. a gland that secretes digestive enzymes, insulin, and glucagon
18. a corticosteroid hormone secreted by the adrenal cortex
19. any one of the many circulating chemical signals secreted by endocrine glands
21. the family of steroids synthesized by and released from the adrenal cortex
22. a hormone produced in the brain and anterior pituitary that inhibits pain perception
24. the abbreviation for the hormone that stimulates the production of eggs and sperm
25. the internal system of the body that uses hormones for chemical communication
26. part of a gland under the hypothalamus, it consists of endocrine cells that secrete hormones directly into blood

27. a hormone produced as a response to stress, also called adrenaline
28. a serious hormonal disease in which the body cells are unable to absorb glucose from the blood
30. the central portion of the adrenal gland
31. the type of hormones that counteract the effects of another hormone

DOWN

2. the outer portion of the adrenal gland
3. the abbreviation for the hormone that raises blood calcium levels
4. the endocrine gland at the base of the hypothalamus
7. clusters of specialized pancreatic cells that produce insulin
8. the type of endocrine gland located adjacent to the kidneys in mammals
9. a hormone produced in response to stress
12. an extension of the hypothalamus composed of nervous tissue that secretes hormones made in the hypothalamus
13. an antagonistic hormone to insulin
14. the primary female steroid sex hormones
20. the type of ductless glands that secrete hormones
21. a mammalian thyroid hormone that lowers blood calcium levels
23. the principal male steroid hormones
29. the abbreviation for the protein hormone that stimulates ovulation in females and androgen production in males
32. the abbreviation for the hormone that stimulates development, growth, and metabolism

CHAPTER 26

Reproduction and Development

Studying Advice

a. Few systems of the body capture our interest and attention more than the one we use to reproduce. This chapter is your chance to learn more about how our reproductive systems function. The topics range from the tragedy of sexually transmitted disease to the joy of having children.

b. This chapter is filled with details of the structures and functions of the reproductive system. Create a system of note cards or other quick question-and-answer review. Create your note cards as you read through the chapter. Then review them daily to master the long list.

Student Media

Activities

Reproductive System of the Human Female
Reproductive System of the Human Male
Sea Urchin Embryonic Development
Frog Embryonic Development

Case Studies in the Process of Science

What Might Obstruct the Male Urethra?
What Determines Cell Differentiation in the Sea Urchin?

Videos

Frog Embryo Development
Rotifer
Sea Urchin Embryonic Development (time-lapse)

MP3 Tutors

Female Reproductive Cycle

Organizing Tables

Compare the definitions of the forms of asexual reproduction in the table below. Provide at least one example of each.

TABLE 26.1		
	Definition	Example
Binary fission		
Fission		
Fragmentation		
Regeneration		
Budding		

Compare oogenesis and spermatogenesis in the table below.

TABLE 26.2		
	Oogenesis	Spermatogenesis
Where each process occurs		
Number of gametes produced from each parent cell		
The distribution of cytoplasm in the cells produced by this process		
The relative size, motility, and nutrient storage in the gametes produced		
Timing of completion of the process		

Identify the location where each of these processes occurs in humans.

TABLE 26.3

	Location
Sperm storage	
Fertilization	
Implantation	

Describe the structure and location of each of the developmental stages in the table below.

TABLE 26.4

	Structure	Location in the Mother's Reproductive Tract
Cleavage		
Blastocyst		
Gastrula		

Content Quiz

Directions: Identify the *one* best answer for the multiple-choice questions. For true/false questions, determine if the statement is true or false. If false, change the underlined word(s) to make the statement true. Finally, add the correct word(s) to the fill-in-the-blank questions to make the statements true.

Biology and Society: Rise of the Supertwins

1. Which one of the following statements about fertility drugs is *false*?

   A. Fertility drugs have allowed thousands of infertile couples to have a baby.

   B. About 30% of women taking fertility drugs become pregnant with more than one embryo.

   C. Increased use of fertility drugs has increased the multiple birth rate.

   D. Fertility drugs stimulate the ovaries to release one or more eggs.

2. Which one of the following statements is *false*? Newborns from multiple pregnancies:

  A. are generally less healthy than newborns from single pregnancies.

  B. have higher birth weights.

  C. are less likely to survive.

  D. are more likely to suffer lifelong disabilities if they survive.

3. True or False? Fertility drugs stimulate the ovaries to release one or more <u>eggs</u>.

4. Increased use of _____ has caused the multiple birth rate in the United States to soar.

Unifying Concepts of Animal Reproduction
ASEXUAL REPRODUCTION

Matching: Match the processes on the left to their best description on the right.

_____ 5. binary fission

_____ 6. fission

_____ 7. fragmentation

_____ 8. budding

  A. breaking up a parent body into pieces followed by regeneration

  B. a single cell splits via mitosis into two genetically identical cells

  C. splitting off new individuals from existing ones

  D. one individual splits into two or more about equal in size

9. Which one of the following statements is *false*? Asexual reproduction:

  A. makes it easier to reproduce if an organism is sessile.

  B. makes it easier to reproduce if organisms are greatly isolated from one another.

  C. takes longer than sexual reproduction.

  D. allows an individual very well suited to its environment to quickly multiply and exploit available resources.

  E. eliminates the need to find a mate.

10. True or False? Asexual reproduction produces genetically <u>diverse</u> populations.

11. The creation of offspring that are genetically identical to a lone parent defines _____ reproduction.

12. The splitting off of a new individual from existing ones is a form of asexual reproduction called _____.

13. The regrowth of a whole animal from pieces is called _____.

SEXUAL REPRODUCTION

14. Compared to asexual reproduction, sexual reproduction is generally more adaptive when:
    A. there is environmental stability.
    B. conditions favor the production of the greatest number of offspring.
    C. conditions favor the production of genetically identical offspring.
    D. the environment changes significantly.
    E. More than one of the above.

15. True or False? Sexual reproduction creates genetically <u>unique</u> offspring through the blending of the genotypes of two parents.

16. True or False? The process of <u>external</u> fertilization occurs when sperm are deposited in or near the female reproductive tract and the gametes fuse within the female's body.

17. The male gamete is the _____ and the female gamete is the _____.

18. Two haploid sex cells unite during fertilization to form a diploid _____.

19. Some species are _____ possessing male and female reproductive systems.

Human Reproduction
FEMALE REPRODUCTIVE ANATOMY

Matching: Match the structures on the left to their best description on the right.

_____ 20. cervix   A. narrow neck at the bottom of the uterus

_____ 21. follicles   B. the site of gamete production in females

_____ 22. oviduct   C. a short, sensitive shaft supporting a rounded glans

_____ 23. ovaries   D. a single egg surrounded by one or more layers of cells

_____ 24. vulva   E. birth canal

_____ 25. uterus   F. common site of fertilization

_____ 26. vagina   G. the site of pregnancy

_____ 27. clitoris   H. collective name for the female reproductive anatomy

28. Which one of the following statements is *false*?
    A. During a menstrual cycle, a woman typically releases one egg cell about every 28 days.
    B. The uterus is about the size and shape of a pear.
    C. After about the ninth week of development, the embryo is called a fetus.
    D. During sexual arousal, the vagina, labia minora, and clitoris engorge with blood.
    E. The uterus is lined with a blood-rich layer of tissue called the myometrium.

29. True or False? It is recommended that women have their cervix examined yearly via a Pap test.

30. The organs that produce gametes are called _____.

31. An egg cell is ejected from the follicle in the process of _____.

MALE REPRODUCTIVE ANATOMY

Matching: Match the structures on the left to their best description on the right.

_____ 32.	penis	A. a tube that stores sperm while they develop
_____ 33.	testes	B. one of the glands producing the fluid portion of semen
_____ 34.	urethra	C. the site of gamete production in males
_____ 35.	epididymis	
_____ 36.	scrotum	D. a sac containing the testes
_____ 37.	prostate	E. a shaft that supports the glans
_____ 38.	vas deferens	F. a tube that conducts sperm and urine
		G. a duct sperm travel through during ejaculation

39. Which one of the following statements is *false*?
    A. About 95% of semen consists of sperm.
    B. The prostate gland is commonly diseased in men over 40.
    C. The testes function best at below normal body temperature.
    D. The seminal vesicle adds part of the fluid that forms semen.
    E. Prostate cancer is the second most commonly diagnosed cancer in the United States.

40. True or False? The production of sperm is greatest at a temperature slightly above normal body temperature.

41. The expulsion of sperm-containing fluid from the penis is called _____.

GAMETOGENESIS

42. Which one of the following statements is *false*?
    A. Meiosis in oogenesis yields cells of unequal size.
    B. Meiosis in spermatogenesis yields cells of equal size.
    C. Meiosis I of spermatogenesis produces four secondary spermatocytes.
    D. Meiosis in human males occurs only within seminiferous tubules.
    E. Meiosis in females starts before birth.

43. Which one of the following statements is *false*?
    A. Polar bodies are produced during oogenesis but not spermatogenesis.
    B. Spermatogenesis produces four gametes but oogenesis results in only one gamete from each parent cell.
    C. Sperm are small, motile, and contain few nutrients, while eggs are large, nonmotile, and well stocked with nutrients.
    D. Human females create mature ova only during fetal development.
    E. Human males create new sperm every day from puberty through old age.

44. True or False? A secondary oocyte completes meiosis II after <u>fertilization</u>.

45. The creation of gametes within the gonads defines _____.

46. At birth, a baby girl has thousands of _____, diploid cells that have paused in prophase of meiosis I.

47. Sperm develop within _____ in the testes.

48. Prior to ejaculation, sperm are stored within the _____.

THE FEMALE REPRODUCTIVE CYCLE

49. Which one of the following statements is *false*?
    A. The menstrual discharge consists of blood, clusters of endometrial cells, and mucus.
    B. The first day of menstruation is designated as the last day of the menstrual cycle.
    C. The endometrium will not be discharged if an embryo implants.
    D. The menstrual discharge leaves the body through the vagina.
    E. After menstruation, the endometrium regrows, reaching its maximum thickness in 20–25 days.

50. True or False? The <u>menstrual</u> cycle controls the growth and release of an ovum.

51. True or False? The growth of an ovarian follicle is stimulated by <u>LH</u>.

52. Uterine bleeding caused by the breakdown of the endometrium defines _____.

53. Ovulation typically takes place on day _____ of the 28-day cycle.

54. FSH and LH are produced by the _____.

55. A positive pregnancy test detects the hormone _____.

Reproductive Health
CONTRACEPTION

Matching: Match the items on the left to their best description on the right.

_____ 56. tubal ligation

_____ 57. vasectomy

_____ 58. rhythm method

_____ 59. withdrawal

_____ 60. diaphragm

_____ 61. cervical cap

_____ 62. condom

_____ 63. spermicides

A. removing the penis from the vagina before ejaculation

B. removal of a short section from each vas deferens

C. a thimble-shaped, small cap that tightly covers the cervix

D. removal of a short section from each oviduct

E. sperm-killing chemicals

F. refraining from intercourse around the time of ovulation

G. a large dome-shaped rubber cap that covers the cervix

H. a barrier that fits over the penis or within the vagina

64. Which one of the following statements is *false*?
   A. The most widely used birth control pills contain synthetic estrogen and progestin.
   B. "The pill" prevents ovulation and keeps follicles from developing.
   C. Combined hormone contraceptives are also available as a shot, a ring inserted into the vagina, or a patch.
   D. Extensive evidence links the pill to cancers.
   E. Birth control pills require a physician's examination and prescription.

65. True or False? There is <u>no chemical</u> contraceptive currently available that prevents the production or release of sperm.

66. If pregnancy has already occurred, the drug _____ can be used to induce an abortion during the first seven weeks of pregnancy.

SEXUALLY TRANSMITTED DISEASES

67. Which one of the following statements is *false*?
   A. AIDS is caused by HIV.
   B. Very few STDs cause long-term problems or death if left untreated.
   C. STDs are most prevalent among teenagers and young adults.
   D. The best way to avoid both unwanted pregnancy and the spread of STDs is abstinence.
   E. Latex condoms provide the best dual protection for "safe sex."

68. True or False? Viral STDs <u>are not</u> curable.

69. True or False? Many people infected with sexually transmitted diseases have <u>no apparent</u> symptoms early in the infection.

70. Chlamydia, gonorrhea, and syphilis are caused by _____.

71. Yeast infections are caused by _____.

72. Genital herpes and genital warts are caused by _____.

Human Development

FERTILIZATION

73. Which one of the following does *not* occur during fertilization?
    A. Fusion of egg and sperm changes the egg so that other sperm cannot penetrate it.
    B. Fructose from the semen fuels movement of the sperm's tail.
    C. Fusion of egg and sperm activates the egg's metabolic machinery.
    D. The sperm penetrates the jelly coat around the egg.
    E. Additional sperm surrounding the egg trigger the egg to undergo mitosis.

74. True or False? The sperm's thick head contains a <u>diploid</u> nucleus.

75. The enzymes that digest a hole in the jelly coat around the egg are found in a sperm's _____.

76. The movement of the sperm tail is powered by _____, organelles clustered near the middle of the sperm.

BASIC CONCEPTS OF EMBRYONIC DEVELOPMENT

Matching: Match each item on the left to its best description on the right.

_____ 77. endoderm
_____ 78. ectoderm
_____ 79. mesoderm
_____ 80. induction
_____ 81. blastocyst
_____ 82. gastrula
_____ 83. cleavage
_____ 84. programmed cell death

A. gives rise to the nervous system
B. results in an embryo shaped like a solid multicellular ball
C. gives rise to the digestive system
D. a hollow ball with three distinct tissue layers
E. a process used to carve out fingers and toes
F. gives rise to the heart, kidneys, and muscles
G. a way that cells influence an adjacent group of cells
H. a fluid-filled hollow ball of about 100 cells

85. Which one of the following does not occur during cleavage?
   A. DNA replication
   B. cytokinesis
   C. an increase in the size of the embryo
   D. mitosis
   E. All of the above do occur during cleavage.

86. True or False? Changes in <u>cell shape</u> help to create embryonic structures.

PREGNANCY AND EARLY DEVELOPMENT

Matching: Match the structures on the left to their best description on the right.

_____ 87. yolk sac   A. it produces the embryo's first blood cells and its first germ cells

_____ 88. allantois   B. a fluid-filled sac that encloses and protects the embryo

_____ 89. amnion   C. forms part of the umbilical cord

_____ 90. chorion   D. becomes part of the placenta

91. Which one of the following statements about the placenta is *false*?
   A. The embryonic and maternal blood supplies in the placenta flow into each other.
   B. Nutrients and oxygen are extracted from the mother's blood.
   C. Embryonic wastes are released into the mother's blood.
   D. Protective antibodies pass from the mother to the fetus.
   E. Most drugs can cross the placenta and harm the embryo.

92. True or False? Most viruses <u>cannot</u> cross the placenta and cause disease.

93. The outer cell layer of the gastrula, the _____, becomes part of the placenta.

94. The _____ is the organ that provides nourishment and oxygen to the embryo and helps dispose of its metabolic wastes.

95. The inner cell mass contains _____ cells, which have the potential to give rise to every type of cell in the body.

THE STAGES OF PREGNANCY

96. Which one of the following does not occur during the third trimester?
   A. The fetus gains the ability to maintain its own temperature.
   B. The fetus's bones begin to harden and muscles thicken.
   C. The fetus loses much of its fine body hair, except on its head.
   D. The fetus rotates so that its head points upward towards the mother's lungs.
   E. All of the above occur during the third trimester.

97. True or False? By the end of the <u>first</u> trimester, the fetus's eyes are open and its teeth are forming.

98. During the _____ trimester, the fetus's circulatory system and respiratory system undergo changes that will permit the breathing of air.

99. All of a fetus's organs and major body parts are formed by the _____ trimester.

100. The most dramatic changes in a fetus occur during the _____ trimester.

CHILDBIRTH

101. Which one of the following statements about oxytocin is *false*?
    A. Oxytocin is a powerful stimulant for the smooth muscles in the wall of the uterus.
    B. Oxytocin stimulates the placenta to make prostaglandins.
    C. Oxytocin and prostaglandins cause uterine contractions.
    D. Estrogen triggers the formation of numerous oxytocin receptors on the uterus.
    E. Oxytocin promotes milk production by the mammary glands.

102. The pituitary hormone <u>estrogen</u> promotes milk production by the mammary glands.

103. The _____ stage of labor is characterized by the birth of the child.

104. A child is born after a series of strong, rhythmic contractions of the uterus called _____.

105. During the final stage of labor, the _____ is delivered.

106. The longest stage of labor is _____, lasting 6–12 hours or longer.

Reproductive Technologies
INFERTILITY

107. Infertility can be due to:
    A. low sperm counts.
    B. inability of sperm to swim to an egg.
    C. a lack of ova.
    D. failure to ovulate.
    E. All of the above.

108. Which one of the following statements is *false?*
    A. Low sperm counts may be due to a scrotum kept too warm by tight-fitting underwear.
    B. Drug therapies (including Viagra®) and penile implants can be used to treat sterility.
    C. Hormone injections can induce ovulation but may result in multiple pregnancies.
    D. As with sperm, ova can be obtained from donors willing to help infertile couples.
    E. A woman able to become pregnant but unable to support a growing fetus might hire a surrogate mother.

109. True or False? Infertility is usually due to problems with the <u>woman</u>.

110. Infertility can be caused by _____, the inability to maintain an erection.

111. Temporary _____ can result from alcohol or drug use, or because of psychological reasons.

IN VITRO FERTILIZATION

112. Which one of the following statements is *false? In vitro* fertilization:
    A. costs around $10,000 for each attempt, successful or not.
    B. begins with the surgical removal of ova and the collection of sperm.
    C. relies on the complete development of a fetus outside the uterus.
    D. permits genetic testing of 8-cell embryos prior to implantation.
    E. is almost routine today.

113. True or False? By implanting only embryos of a certain sex, *in vitro* fertilization <u>can be used</u> by couples to select the sex of their baby.

114. "*In vitro*" literally means _____.

Evolution Connection: Menopause and the Grandmother Hypothesis

115. Anthropologist Kristen Hawkes and her colleagues found that in a tribe of hunter-gatherers in northern Tanzania, children with caring grandmothers:
    A. gained more weight and grew faster.
    B. answered math problems more quickly.
    C. had the best artistic talents.
    D. were generally overweight.
    E. usually did not have living parents.

116. True or False? Most species <u>lose</u> their reproductive capacity throughout life.

117. The cessation of ovulation and menstruation is called _____.

Word Roots

a = without (asexual reproduction: reproduction without sex)

bi = two (binary fission: division into two new organisms)

blasto = produce; **cyst** = sac, bladder (blastocyst: a hollow ball of cells produced one week after fertilization in humans)

contra = against (contraception: the prevention of pregnancy)

ecto = outer; **derm** = tissue (ectoderm: the outermost tissue layer of an embryo)

ectomy = cut out (vasectomy: the cutting of each vas deferens to prevent sperm from entering the urethra)

endo = inner (endoderm: the innermost tissue layer of an embryo)

endo = inside (endometrium: the inner lining of the uterus, which is richly supplied with blood vessels)

epi = above, over (epididymis: a coiled tubule located adjacent to the testes where sperm are stored)

fertil = fruitful (fertilization: the union of haploid gametes to produce a diploid zygote)

gastro = stomach, belly (gastrulation: the formation of a gastrula from a blastula)

in = without (infertility: the inability to reproduce sexually)

labi = lip; **major** = larger (labia majora: a pair of thick, fatty ridges that enclose and protect the labia minora and vestibule)

meso = middle (mesoderm: the middle tissue layer of an embryo)

oo = egg; **genesis** = producing (oogenesis: the process in the ovary that results in the production of female gametes)

ovi = egg (oviduct: the tube that carriers an egg from the ovary)

re = again (regeneration: the replacement of a body part)

tri = three (trimester: a three-month period)

tropho = nourish (trophoblast: the outer epithelium of the blastocyst, which forms the fetal part of the placenta)

Key Terms

acrosome
allantois
amnion
asexual
  reproduction
barrier methods
binary fission
birth control pills
blastocyst
cervix
chorion
chorionic villi
cleavage
clitoris
contraception
copulation
corpus luteum
diaphragm
ectoderm
ejaculation
embryo
endoderm
endometrium
epididymis
external
  fertilization
fertilization
fetus
fission
follicles

fragmentation
gametes
gametogenesis
gastrula
gastrulation
gestation
glans
gonads
hermaphrodite
hymen
impotence
in vitro
  fertilization
  (IVF)
induction
infertility
inner cell mass
internal
  fertilization
labia majora
labia minora
labor
menstrual cycle
menstruation
mesoderm
morning after pills
  (MAPs)
natural family
  planning
oogenesis

ovarian cycle
ovaries
oviduct
ovulation
ovum
oxytocin
penis
placenta
polar body
prepuce
primary oocyte
primary
  spermatocytes
programmed cell
  death
prostate
regeneration
reproduction
reproductive cycle
rhythm method
scrotum
secondary oocyte
secondary
  spermatocytes
semen
seminal vesicles

seminiferous
  tubules
sexual
  reproduction
sexually
  transmitted
  diseases (STDs)
sperm
spermatogenesis
spermicides
stem cells
testes
trimesters
trophoblast
tubal ligation
umbilical cord
urethra
uterus
vagina
vas deferens
vasectomy
vulva
withdrawal
yolk sac
zygote

Crossword Puzzle

Use the Key Terms list from this chapter to fill in the crossword puzzle.

ACROSS

1. a female organ in mammals where embryos develop
3. a gland that secretes a fluid component of semen that lubricates and nourishes sperm
4. in human development, one of three 3-month-long periods of pregnancy
6. the collective name for the outer features of the female reproductive anatomy
8. the regrowth of body parts from pieces of an organism
9. a form of contraception also known as natural family planning
11. the remainder of a follicle after ovulation
12. a pouch of skin outside the abdomen that houses the testes
13. the type of reproduction resulting from the joining of parental genetic material
14. the prevention of pregnancy
19. the abbreviation for contagious diseases spread by sexual contact
20. the creation of new individuals from existing ones
22. a sensitive female sexual organ that becomes erect during sexual arousal
23. the ability of one group of embryonic cells to influence the development of another
24. sexual intercourse
25. a fold of skin covering the head of the clitoris or penis
26. the process of cytokinesis in animal cells
27. a type of spermatocyte produced by meiosis I
28. the type of fertilization in which the gametes fuse outside the parents' bodies
30. the inability to maintain an erection
32. the extraembryonic membrane that contributes to the formation of the mammalian placenta

33. a male gamete
34. the neck of the uterus that opens into the vagina
36. the inner lining of the uterus
37. removal of the penis from the vagina before ejaculation
38. chemical contraceptives taken orally that are very effective at preventing pregnancy
39. the cutting of each vas deferens to prevent sperm from entering the urethra
40. the embryonic stage consisting of a hollow ball of cells
41. the innermost of the three primary germ layers in animal embryos
45. an extraembryonic membrane that forms a fluid-filled embryonic sac
46. the inability to conceive a child
47. contraception that relies upon a physical barrier
49. the type of spermatocyte that undergoes meiosis I
50. an extraembryonic membrane that stores embryonic nitrogenous waste
52. the cyclic recurrence of the follicular phase, ovulation, and the luteal phase in the mammalian ovary
55. the process that produces sperm
57. the hormone that induces contractions of uterine muscles and ejection of milk during nursing
59. in mammals, the tube passing from the ovary to the uterus
60. a dome-shaped rubber cap that covers the cervix
62. the female gamete
65. the type of fertilization inside a parent's body
68. a haploid cell such as an egg or sperm

69. a structure in the pregnant uterus for nourishing a viviparous fetus with the mother's blood supply

71. the female reproductive gonad

74. a coiled tubule that stores sperm in males

75. a type of cell that gives rise to other cells and restores its own population

76. the process that forms a gastrula

77. the type of cycle in higher female primates that leads to a bloody discharge

78. the male and female sex organs

79. the abbreviation for the process of fertilizing ova in laboratory containers

81. the abbreviation for a type of pill used as emergency contraception

83. an individual that produces sperm and eggs

84. the three-layered, cup-shaped embryonic stage in animals

85. the production of female gametes

86. the head of the shaft of the penis

87. a means of asexual reproduction by which one individual separates into two or more individuals of about equal size

88. a means of sterilization in which a woman's oviducts are tied closed

89. a means of asexual reproduction whereby a single parent breaks into parts that regenerate into whole new individuals

DOWN

1. a tube that drains urine from the urinary bladder

2. the fluid that is ejaculated by a male during orgasm

4. the male reproductive gonad

5. a type of male gland that secretes an acid-neutralizing component of semen

6. the birth canal in mammals

7. a recurring series of events that produce gametes in females

10. an extraembryonic membrane that stores yolk in bird and reptile eggs

11. the knobby outgrowths on the outside of the chorion

13. a type of haploid oocyte resulting from meiosis I in oogenesis, which will become an ovum after meiosis II

15. the outermost of the three primary germ layers in animal embryos

16. the type of highly coiled tubules where sperm are produced

17. the outer epithelium of the blastocyst, which forms the fetal part of the placenta

18. the lifeline between the embryo and the placenta

21. another name for the rhythm method

29. the type of cell division used by prokaryotes to reproduce by splitting into two

31. the copulatory structure of male mammals

33. sperm-killing chemicals

35. the release of an egg from ovaries

36. a new developing individual

42. a cell produced during meiosis that does not develop further or participate in fertilization

43. the middle of the three primary germ layers in animal embryos

44. a thin membrane that partly covers the vaginal opening in a human female

45. an organelle at the tip of a sperm that helps it penetrate an egg

48. the type of oocyte that is a diploid cell, often arrested in prophase I of meiosis

51. a pair of thin ridges that enclose the vestibule

53. the type of reproduction involving only one parent

54. the production of eggs or sperm

56. the shedding of portions of the endometrium during a menstrual cycle

58. the expulsion of sperm-containing fluid from the penis

61. the portion of a blastocyst that will form the baby

63. a pair of thick, fatty ridges that enclose and protect the labia minora and vestibule

64. a type of cell death

66. rhythmic contractions of the uterus that give rise to the birth of a child

67. the union of haploid gametes to produce a diploid zygote

70. the tube that transports sperm from the epididymis during ejaculation

72. the state of carrying developing young within the female reproductive tract

73. a developing human from the ninth week of gestation until birth

80. a microscopic structure in an ovary that contains the developing ovum and secretes estrogens

82. the diploid product of fertilization

CHAPTER 27

Nervous, Sensory, and Motor Systems

Studying Advice

a. Have you ever thought about how you think? Ever wondered how your brain keeps everything sorted out? Ever consider how you remember anything? This fascinating chapter reveals insights into many of these amazing processes.

b. This is not a chapter for one night of studying. It is long and it contains an extensive list of vocabulary terms. The organizing tables below help you sort out some of the information, but you will need to create other systems to learn the vocabulary. Consider using the list of key terms towards the end of this chapter to create note cards with the terms on one side and the definitions on the other. You will need to quiz yourself over many days to master all of these.

Student Media

Activities

Neuron Structure

Nerve Signals: Action Potentials

Neuron Communication

Structure and Function of the Eye

The Human Skeleton

Skeletal Muscle Structure

Muscle Contraction

Case Studies in the Process of Science

What Triggers Nerve Impulses?

How Do Electrical Stimuli Affect Muscle Contraction?

eTutors

How Neurons Work

Videos

Discovery Channel Video Clip: Teen Brains

Discovery Channel Video Clip: Muscles and Bones

Organizing Tables

Compare the two main subdivisions of the nervous system in the table below. In the function(s) category, indicate which systems are responsible for sensory input, integration, and motor output.

TABLE 27.1

	Structural Components	Function(s)
CNS		
PNS		

Compare the structures and functions of the parts of a neuron in the table below.

TABLE 27.2

	Structural	Function(s)
Cell body		
Dendrite		
Axon		

Compare the structures and functions of the components of the peripheral nervous system in the table below.

TABLE 27.3		
	Structural Components	Function(s)
Sensory division		
Motor division: Somatic nervous system		
Motor division: Autonomic nervous system, sympathetic division		
Motor division: Autonomic nervous system, parasympathetic division		

Compare the five general categories of sensory receptors in the table below.

TABLE 27.4		
	Location	Function(s)
Pain receptors		
Thermoreceptors		
Mechanoreceptors		
Chemoreceptors		
Electromagnetic receptors		

Content Quiz

Directions: Identify the *one* best answer for the multiple-choice questions. For true/false questions, determine if the statement is true or false. If false, change the underlined word(s) to make the statement true. Finally, add the correct word(s) to the fill-in-the-blank questions to make the statements true.

Biology and Society: Treating Depression

1. Which one of the following is *not* a characteristic of depression?
    A. persistent sadness
    B. a loss of interest in pleasurable activities
    C. changes in sleeping patterns
    D. changes in body weight
    E. increased energy

2. True or False?   Once thought to be a purely psychological condition, depression is now known to be <u>a disorder of the brain</u>.

3. Depression is frequently associated with a lower level of one particular neurotransmitter, called _____.

An Overview of Animal Nervous Systems
ORGANIZATION OF NERVOUS SYSTEMS, NEURONS

4. If you touch something sharp and pull your finger back to avoid injury, the signals travel from:
    A. sensory input to integration to motor output.
    B. sensory input to motor output to integration.
    C. integration to sensory input to motor output.
    D. motor output to integration to sensory input.
    E. motor output to sensory input to integration.

5. True or False?   Signals from the central nervous system are sent out to effectors by <u>sensory</u> neurons.

6. True or False?   The <u>central</u> nervous system is mostly composed of nerves that carry signals into and out of the CNS.

7. A(n) _____ is a communication line made from cable-like bundles of neuron fibers tightly wrapped in connective tissue.

8. The conveyance of signals to the CNS from sensory receptors is called _____.

9. The part of a neuron that houses the nucleus is called the _____.

10. Outnumbering neurons are _____ cells that protect, insulate, and reinforce the neurons.

11. Axons that convey signals very rapidly are enclosed along most of their length by an insulating material called the _____.

12. Signals are carried toward another neuron or toward an effector by _____.

13. Gaps in the myelin sheath form _____, the only points on the axon where signals can be transmitted.

SENDING A SIGNAL THROUGH A NEURON

14. Stimulating a neuron's plasma membrane to generate a nerve signal is most like:
    A. hitting a baseball with a bat.
    B. turning on a flashlight to create light.
    C. bouncing a ball against the ground.
    D. cutting up paper with scissors.
    E. constructing a dog house.

15. Which one of the following statements about membrane potentials is *false*?
    A. The potential energy of a cell is in an electrical charge difference across a neuron's plasma membrane.
    B. During a resting membrane potential, the cytoplasm just inside the membrane is negative in charge.
    C. A membrane stores energy by joining opposite charges together.
    D. A cell's membrane has channels and pumps that regulate the passage of positive ions, contributing to the resting membrane potential.
    E. During a resting potential, the fluid just outside the cell is positive.

16. How do the events of an action potential cause the "domino effect" of a nerve signal?
    A. No ions are allowed to pass through the membrane where an action potential occurs so ions cross at the next available point.
    B. Inflowing negative ions trigger the opening of channels in the membrane next to the action potential.
    C. Inflowing positive ions trigger the opening of channels in the membrane next to the action potential.
    D. Changes in the myelin sheath create new openings in regions next to the place where an action potential is occurring.

17. How do action potentials relay different intensities of information to the central nervous system?

    A. A stronger action potential is sent when a signal is stronger.

    B. A weaker action potential is sent when a signal is stronger.

    C. Stronger signals cause a greater surge of ions across the membrane.

    D. The frequency of action potentials changes with the intensity of stimuli.

18. True or False?   Action potentials are <u>all-or-none</u> events.

19. True or False?   Most of the dissolved proteins and other large organic molecules inside a neuron are <u>positively</u> charged.

20. The voltage across a plasma membrane of a resting neuron is called the _____.

21. Any factor that causes a nerve signal to be generated is called a(n) _____.

22. If a stimulus is strong enough, a sufficient number of channels open to reach the _____, the minimum change in a membrane's voltage that must occur to trigger an action potential.

PASSING A SIGNAL FROM A NEURON TO A RECEIVING CELL

Matching: Match each item on the left to its best description on the right.

_____ 23. caffeine

_____ 24. nicotine

_____ 25. alcohol

_____ 26. opiates

_____ 27. antipsychotic drugs

_____ 28. ritalin

_____ 29. endorphins

_____ 30. Parkinson disease

_____ 31. schizophrenia

A. decrease our perception of pain

B. chemically similar to dopamine and norepinephrine

C. a strong depressant, it seems to increase the effects of GABA

D. a disease associated with a lack of dopamine

E. binds to and activates receptors for acetylcholine

F. counters the effects of inhibitory neurotransmitters

G. a disease associated with an excess of dopamine

H. drugs that bind to endorphin receptors

I. block dopamine receptors

32. Where are neurotransmitters located before an action potential travels down a nerve?

    A. in the synaptic cleft

    B. in the synaptic knob

    C. in the membrane of the neuron

    D. in the dendrites

    E. in the nerve cell body

33. What happens to neurotransmitters after they bind to a receiving neuron's plasma membrane?

    A. They are absorbed by the receiving neuron's plasma membrane.

    B. They are sent to the nerve cell body.

    C. They travel down the receiving neuron's plasma membrane.

    D. They are broken down or transported back to the sending neuron for recycling.

34. True or False? Synapses <u>can be</u> either electrical or chemical.

35. True or False? In <u>a chemical</u> synapse, an action potential jumps directly from one neuron to the next.

36. The narrow gap separating a synaptic knob of the sending neuron from the receiving neuron is called a(n) _____.

37. A relay point between two neurons or between a neuron and an effector cell is a(n) _____.

38. In a(n) _____ synapse, an action potential is converted to chemical signals.

39. A(n) _____ carries information from one nerve cell to another in a chemical synapse.

The Human Nervous System: A Closer Look
THE CENTRAL NERVOUS SYSTEM

40. Which one of the following statements about the spinal cord is *false?*

    A. The spinal cord functions like a telephone cable jam-packed with wires.

    B. The spinal cord is often severed.

    C. The spinal cord contains white matter.

    D. The spinal cord contains gray matter.

    E. Meninges protect the brain and spinal cord.

41. True or False? <u>White</u> matter mainly consists of axons with light-colored myelin sheaths.

42. True or False? The concentration of the nervous system at the head end of the body is called <u>centralization</u>.

43. True or False? Paralysis of the lower half of the body is called <u>quadriplegia</u>.

44. The presence of a central nervous system distinct from a peripheral nervous system is called _____.

45. The _____ is the master control center of the nervous system.

46. An infection of the meninges is called _____.

THE PERIPHERAL NERVOUS SYSTEM

47. Which one of the following does *not* occur when the sympathetic division of the autonomic nervous system is activated?
    A. pupils dilate
    B. heart accelerates
    C. glucose is released into the bloodstream
    D. epinephrine and norepinephrine are released into the bloodstream
    E. salivary production is increased

48. True or False? The <u>autonomic</u> nervous system is said to be voluntary, because most of its actions are under conscious control.

49. True or False? The <u>sympathetic</u> division primes the body for digesting food and resting.

50. True or False? Neurons of the <u>autonomic</u> nervous system carry signals to skeletal muscles.

51. If you suddenly realize that you missed a lecture exam, your _____ division of your autonomic nervous system will likely respond.

THE HUMAN BRAIN

Matching: Match each structure on the left to its best description on the right.

_____ 52. thalamus	A. the highly folded outer surface of the cerebrum
_____ 53. hypothalamus	B. consists of the medulla oblongata and pons
_____ 54. brainstem	C. regulates body temperature, blood pressure, hunger, and thirst
_____ 55. cerebellum	
_____ 56. corpus callosum	D. consists of the thalamus, hypothalamus, and cerebral cortex
_____ 57. cerebral cortex	
_____ 58. limbic system	E. provides coordination of movement and balance
	F. a band that connects the cerebral hemispheres
	G. sorts data into categories; suppresses or enhances other signals

59. Which one of the following statements is *false?*
    A. The cerebral cortex is divided into right and left sides.
    B. The corpus callosum restricts communication between the two hemispheres.
    C. Each side of the cerebral cortex has four lobes.
    D. The association areas are the sites of higher mental activities.
    E. Language results from extremely complex interactions among several association areas.

60. True or False?   The <u>medulla oblongata</u> is a planning center for body movements.

61. Areas in the two cerebral hemispheres become specialized for different functions in the process of _____.

62. One part of the hypothalamus functions as a timing mechanism, a sort of _____ clock.

The Senses
SENSORY INPUT

Matching: Match each structure on the left to its best description on the right.

_____ 63. pain receptors	A.	detect various forms of mechanical energy
_____ 64. thermoreceptors	B.	respond to chemicals
_____ 65. mechanoreceptors	C.	respond to energy of various wavelengths
_____ 66. chemoreceptors	D.	detect either heat or cold
_____ 67. electromagnetic receptors	E.	respond to excess heat or pressure

68. Which one of the following receptors is *not* found in skin?
    A. thermoreceptors
    B. electromagnetic receptors
    C. pain receptors
    D. mechanoreceptors
    E. All of the above are found in the skin.

69. True or False?   Receptor cells convert one type of signal (the stimulus) into an electrical signal in a process called sensory <u>adaptation</u>.

70. True or False?   The stronger the stimulus, the <u>stronger</u> the receptor potential.

71. The change in membrane potential in a receptor cell is called the _____.

72. Local regulators that increase pain by sensitizing pain receptors are called _____.

VISION

Matching: Match each structure on the left to its best description on the right.

_____ 73. astigmatism

_____ 74. farsightedness

_____ 75. presbyopia

_____ 76. nearsightedness

_____ 77. cones

_____ 78. rods

_____ 79. lacrimal gland

_____ 80. conjunctiva

_____ 81. glaucoma

_____ 82. fovea

_____ 83. retina

_____ 84. choroid

_____ 85. iris

_____ 86. pupil

_____ 87. sclera

_____ 88. cornea

A. at the retina's center of focus

B. a pigmented layer between the sclera and retina

C. a mucous membrane that helps keep the eye moist

D. a tough, whitish layer of connective tissue

E. secretes a dilute salt solution

F. blurred vision caused by a misshapen lens or cornea

G. the opening in the center of the iris

H. it lets light into the eye and also helps focus light

I. a layer that contains photoreceptor cells

J. a type of farsightedness due to inflexibility of the lens

K. it gives the eye its color

L. detect shades of gray in the retina

M. ability to see far but not near

N. caused by increased pressure inside the eye

O. ability to see near but not far

P. detect color in the retina

89. Which one of the following statements is *false*?

    A. The lens of the eye focuses light onto the retina.

    B. The shape of the lens is controlled by muscles attached to the choroid.

    C. When the eye focuses on a nearby object the lens becomes thinner and flatter.

    D. There are no photoreceptor cells in the part of the retina where the optic nerve passes through the back of the eye.

    E. An infection or allergic reaction may cause inflammation of the conjunctiva, a condition called conjunctivitis.

90. True or False?   The much smaller chamber in front of the lens contains a thin fluid, the <u>vitreous</u> humor.

91. Rods contain a visual pigment called _____, which can absorb dim light. Cones contain a visual pigment called _____.

HEARING

Matching: Match each structure on the left to its best description on the right.

_____ 92. pinna

_____ 93. auditory canal

_____ 94. Eustachian tube

_____ 95. cochlea

_____ 96. organ of Corti

_____ 97. eardrum

A. a tube extending from the pinna to the eardrum

B. the bendable structure we commonly call our "ear"

C. one of the channels in the inner ear

D. an array of hair cells embedded in a basilar membrane

E. conducts air between the middle ear and the back of the throat

F. a sheet of tissue that separates the outer ear from the middle ear

Sequencing. Indicate the correct sequence of the following nine events that occur in the process of hearing. Place a 1 in front of the first event, a 2 in front of the second, etc.

_____ 98. Hair cells develop a receptor potential and release more neurotransmitter molecules at its synapse with a sensory neuron.

_____ 99. As a pressure wave passes through the cochlea, it pushes downward and makes the basilar membrane vibrate.

_____ 100. When a hair cell's projections are bent, ion channels in its plasma membrane open, and positive ions enter the cell.

_____ 101. A vibrating object creates pressure waves in the surrounding air.

_____ 102. Vibrations pass through the hammer, anvil, and stirrup in the middle ear.

_____ 103. The sensory neuron sends more action potentials to the brain through the auditory nerve.

_____ 104. Waves make the eardrum vibrate with the same frequency as the sound.

_____ 105. Vibration of the basilar membrane makes the hairlike projections on the hair cells alternately brush against and draw away from the overlying membrane.

_____ 106. The stirrup transmits the vibrations to the inner ear, producing pressure waves in the fluid within the cochlea.

107. The actual sensory transduction converting sound into an action potential occurs in the:

    A. outer ear.

    B. middle ear.

    C. inner ear.

    D. All of the above.

    E. None of the above.

108. Deafness can be caused by:
   A. the inability to conduct sounds.
   B. a ruptured eardrum.
   C. stiffening of the middle-ear bones.
   D. damage to receptor cells or neurons.
   E. All of the above.

109. True or False?   The three small bones are located in the inner ear.

110. A ringing or buzzing sound in your ears is a condition called _____.

Motor Systems
THE SKELETAL SYSTEM

111. Which one of the following statements about osteoporosis is *false*?
   A. Prevention of osteoporosis begins with sufficient calcium intake while bones are still increasing in density.
   B. Walking, jogging, and lifting weights builds bone mass and is beneficial throughout life.
   C. Increased levels of estrogen contributes to osteoporosis.
   D. Insufficient exercise, an inadequate intake of protein and calcium, smoking, and diabetes mellitus may contribute to osteoporosis.
   E. Treatments include calcium and vitamin supplements, hormone replacement therapy, and drugs that slow bone loss or increase bone formation.

112. Which one of the following is an autoimmune disease in which the joints become highly inflamed and their tissues possibly destroyed?
   A. rheumatoid arthritis
   B. Parkinson disease
   C. tinnitus
   D. osteoporosis
   E. multiple sclerosis

113. True or False?   The axial skeleton is made up of the bones of the limbs, shoulders, and pelvis.

114. True or False?   Humans, like all vertebrates, have an endoskeleton, hard supporting elements situated among soft tissues.

115. True or False?   The shaft of a long bone surrounds a central cavity that contains red bone marrow, mostly stored fat brought into the bone by the blood.

116. A(n) _____ joint allows us to rotate the forearm at the elbow.

117. A freely moving _____ joint joins the humerus to the scapula and the femur to the pelvis.

118. A(n) _____ joint between the humerus and the head of the ulna permits movement in a single plane.

THE MUSCULAR SYSTEM, STIMULUS AND RESPONSE: PUTTING IT ALL TOGETHER

119. Which one of the following does *not* occur during a muscle contraction?
    A. A myosin head gains energy from the breakdown of NADH.
    B. An energized myosin head binds to an exposed binding site on actin.
    C. The molecular event that actually causes sliding is called the power stroke.
    D. During the power stroke, energy is released from the myosin head, and the head bends back to its low-energy position, pulling the thin filament toward the center of the sarcomere.
    E. On the next power stroke, the myosin head attaches to another binding site ahead of the previous one on the thin filament.

120. Which one of the following summarizes the activities of actin and myosin during a muscle contraction?
    A. cock, detach, attach, bend
    B. detach, cock, attach, bend
    C. bend, cock, detach, attach
    D. attach, cock, bend, detach
    E. bend, detach, attach, cock

121. A motor unit functions most like:
    A. blowing into a tuba to make music.
    B. breaking a stick in half.
    C. burning gasoline to make a truck engine function.
    D. a switch that controls a set of lights.
    E. assembling pieces of a puzzle.

122. Which one of the following does *not* occur when a muscle relaxes?
    A. Motor neurons stop sending action potentials to the muscle fibers.
    B. Actin binding sites are blocked again.
    C. The ER releases Ca^{2+} into the cytoplasm.
    D. Sarcomeres stop contracting.

123. Which one of the following is an example of a motor response by a baseball player?

    A. seeing a ball

    B. hearing the crack of the bat against the ball

    C. determining where the ball will go

    D. reaching out to catch the ball

    E. feeling joy in success

124. True or False? The shortening of <u>sarcomeres</u> causes muscles to shorten.

125. True or False? <u>Thin</u> filaments are composed of myosin.

126. True or False? A sarcomere contracts when its thin filaments <u>slide across</u> its thick filaments.

127. Muscles attach to bone by _____.

128. A myofibril consists of repeating units called _____.

129. Each muscle fiber consists of a bundle of smaller _____.

130. A motor neuron and all the muscle fibers it controls define a(n)_____.

131. A motor neuron forms synapses with the muscle fibers at _____ junctions.

Evolution Connection: Convergent Evolution of Eyes

132. Planarians use their eyes to

    A. detect light intensity and direction.

    B. form general images, but only in shades of gray.

    C. form general images in color.

    D. form very specific and clear images, but only in shades of gray.

    E. form very specific and clear images in color.

133. True or False? Planarians tend to move <u>towards</u> a light source.

134. Planarians demonstrate the adaptive advantages of even very simple _____.

Word Roots

aqua = water; **humor** = fluid (aqueous humor: a thin watery fluid in front of the lens of the eye)

arthro = joint; **itis** = inflammation (arthritis: inflammation of the joints of the skeleton)

audi = hear (auditory canal: the tube that channels sound to the eardrum)

auto = self (autonomic nervous system: a subdivision of the motor nervous system of vertebrates that regulates the internal environment)

bi = two (bipolar disorder: a manic-depressive disorder in which a person switches between two extremes)

cephalo = head (cephalization: the clustering of sensory neurons and other nerve cells to form a small brain near the anterior end and mouth of animals with elongated, bilaterally symmetrical bodies)

cerebro = brain (cerebrum: the largest and most sophisticated part of the human brain)

chemo = chemical (chemoreceptors: sensory cells that respond to chemicals)

dendro = tree (dendrite: one of usually numerous, short, highly branched processes of a neuron that conveys nerve impulses toward the cell body)

endo = inner (endoskeleton: a skeleton inside the body)

hypo = below (hypothalamus: the master control center of the endocrine system, located below the thalamus in the brain)

inter = between (interneurons: an association neuron; a nerve cell within the central nervous system that forms synapses with sensory and motor neurons and integrates sensory input and motor output)

myo = muscle (myofibril: a contractile thread in a muscle cell)

neuro = nerve; **trans** = across (neurotransmitter: a chemical messenger released from the synaptic terminal of a neuron at a chemical synapse that diffuses across the synaptic cleft and binds to and stimulates the postsynaptic cell)

osteo = bone; **pori** = a hole (osteoporosis: a disease of thinning and porous bones)

para = near (parasympathetic division: one of two divisions of the autonomic nervous system)

photo = light (photopsin: a light-sensing pigment found in cones)

soma = body (somatic nervous system: a branch of the nervous system that carries signals to skeletal muscles)

syn = together (synapse: the locus where a neuron communicates with a postsynaptic cell in a neural pathway)

thermo = heat (thermoreceptors: sensory receptors that detect heat)

Key Terms

action potential
appendicular skeleton
aqueous humor
arthritis
association areas
astigmatism
auditory canal
autonomic nervous system
axial skeleton
axon
ball-and-socket joints
basilar membrane
biological clock
bipolar disorder
brain
brain stem
cell body
central nervous system (CNS)
centralization
cephalization
cerebellum
cerebral cortex
cerebrospinal fluid
cerebrum
chemoreceptors
choroid
cochlea

cones
conjunctiva
cornea
corpus callosum
dendrites
eardrum
effectors
electromagnetic receptors
endoskeleton
Eustachian tube
farsightedness
fovea
glaucoma
gray matter
hinge joint
hypothalamus
inner ear
interneurons
iris
lacrimal gland
lateralization
lens
ligaments
limbic system
major depression
mechanoreceptors
medulla oblongata
meninges
middle ear
motor neurons

motor units
myelin sheath
myofibrils
nearsightedness
nerve
nervous system
neurons
neurotransmitter
nodes of Ranvier
optic nerve
organ of Corti
osteoporosis
outer ear
pain receptors
parasympathetic division
peripheral nervous system (PNS)
photoreceptors
pinna
pivot joint
pons
pupil
receptor potential
red bone marrow
resting potential
retina
rheumatoid arthritis

rods
sarcomeres
sclera
sensory adaptation
sensory neurons
sensory transduction
skeletal muscle
somatic nervous system
spinal cord
stimulus
supporting cells
sympathetic division
synapse
synaptic cleft
synaptic terminal
tendons
thalamus
thermoreceptors
thick filaments
thin filaments
threshold
vitreous humor
white matter
yellow bone marrow
Z lines

Crossword Puzzle

Use the Key Terms list from this chapter to fill in the crossword puzzle.

ACROSS

1. the type of muscle filament made from actin

2. another name for an association neuron

5. the type of bone marrow that consists mostly of stored fat

6. the type of gland that secretes tears

8. the fundamental unit of muscle contraction

10. the structure that connects the middle ear to the pharynx

14. a hard inner skeleton

16. strong, fibrous tissues that join bones to muscles

17. the region of the vertebrate ear that includes the cochlea, organ of Corti, and semicircular canals

18. the largest and most dominant portion of the vertebrate forebrain

19. strong, fibrous tissues that join bones

20. a thick band of nerve fibers that interconnects the cerebral hemispheres

21. the abbreviation for the brain and spinal cord in vertebrate animals

23. the type of joint that enables mammals to rotate the forearm at the elbow

24. a sophisticated integrating center in the forebrain located just above the hypothalamus

25. the complex, coiled organ of hearing that contains the organ of Corti

26. a short, highly branched processes of a neuron that conveys nerve impulses toward the cell body

29. the transparent frontal portion of the sclera, which admits light into the vertebrate eye

30. vision problem that occurs when the focal point of the lens is behind the retina

31. the presence of a central nervous system distinct from a peripheral nervous system

32. the main eye structure that focuses light onto the retina

33. the type of receptors that respond to electromagnetic energy

34. the type of muscle generally responsible for voluntary movements

35. the master control center of the central nervous system

37. a photoreceptor in the vertebrate eye that detects color

39. the mucous membrane that helps keep the eye moist

40. a receptor of light

41. one of the three main regions of the ear where sound is first collected

43. the opening in the iris

45. the surface of the cerebrum

46. the part of the skeleton associated with the limbs and girdles

48. a thin, pigmented inner layer of the vertebrate eye

50. the type of joint that enables mammals to move arms and legs in several planes

51. layers of connective tissue between the brain and spinal cord

52. in the organ of Corti, which is where hair cells are embedded

53. the portion of a neuron that carries nerve impulses away from the cell body

55. a broad band of nerve fibers that connect the sides of the cerebellum and medulla oblongata

57. inflammation of the joints

58. increased pressure inside the eye that could lead to blindness

59. the type of joint that permits movement in a single plane

60. the type of potential that results from the voltage across the plasma membrane of a resting neuron

61. the type of bone marrow that produces blood cells

62. an internal timekeeper that controls an organism's biological rhythms

63. the type of neurons that transmit signals from the brain or spinal cord to muscles or glands

64. a phenomenon that occurs when the two hemispheres of the brain become specialized for different functions

66. the ends of sarcomeres

67. the sites of higher mental activities

69. the overall feeling of sadness and loss of interest in pleasurable activities

71. the hindbrain and midbrain of the vertebrate central nervous system

72. the part of the skeleton associated with the skull and vertebrae

76. the flaplike part of an ear that initially collects sound

78. the type of system that forms a communication and coordinating network among all parts of an animal's body

79. the part of the vertebrate hindbrain that coordinates movement and balance

80. the type of nerve that carries sensory signals from the eye

81. a photoreceptor in the vertebrate retina that is sensitive to black and white

82. the tube that channels sound waves to the eardrum

83. a relay point between two neurons or between a neuron and an effector cell

84. the abbreviation for the portion of the nervous system that includes sensory and motor neurons

86. the actual hearing organ of the vertebrate ear, located in the inner ear

87. a sheet of tissue that separates the outer ear from the middle ear

88. a type of disorder that is also called manic-depressive disorder

89. a chemical messenger released from the synaptic terminal of a neuron at a chemical synapse

91. a tough, white outer layer of connective tissue that forms the outside of the vertebrate eye

92. the type of receptors that detect injury

93. tracts of axons within the CNS

95. the lowest part of the vertebrate brain that controls autonomic, homeostatic functions

96. the innermost layer of the vertebrate eye containing rods and cones

97. a bone disorder marked by thinner and more easily broken bones

100. the type of cells in the nervous system that protect, insulate, and reinforce a neuron

101. the type of potential that varies with the strength of the stimulus

102. the vision problem that occurs when the focal point of the lens is in front of the retina

103. the dorsal hollow nerve cord in vertebrates

104. an insulating layer around an axon

105. the conversion of a stimulus signal into an electrical signal by a sensory receptor cell

DOWN

1. the type of muscle filament made from myosin

3. a muscle cell or gland cell that performs the body's responses to stimuli

4. a narrow gap separating the synaptic knob of a transmitting neuron from a receiving neuron or an effector cell

7. the colored portion of the choroid at the front of the eye

9. the type of arthritis that is an autoimmune disease

11. a type of fibrous connective tissue that attaches muscle to bone

12. the type of neurons that receive information from sensory receptors

13. regions of dendrites and clusters of nerve-cell bodies within the CNS

15. any factor that causes a nerve signal to be generated

18. blood-derived fluid that surrounds and cushions the brain and spinal cord

21. a sensory structure that responds to chemical changes

22. an eye's center of focus and the place on the retina where photoreceptors are highly concentrated

27. the division of the autonomic nervous system that generally enhances body activities that gain and conserve energy

28. the part of a neuron that houses the nucleus

36. a rapid change in the membrane potential of an excitable cell

38. the minimum change in a membrane's voltage that must occur to trigger an action potential

42. an evolutionary trend toward the concentration of sensory equipment on the anterior end of the body

44. the lower part of the mammalian forebrain that includes the hippocampus and the amygdala

47. plasmalike liquid in the space between the lens and the cornea

49. the ventral part of the vertebrate forebrain that helps maintain homeostasis

54. another name for a nerve cell

56. a contractile thread in a muscle cell made of many sarcomeres

65. the division of the autonomic nervous system that increases energy expenditure and prepares the body for action

68. blurred vision caused by a misshapen lens or cornea

70. a ropelike bundle of neuron fibers tightly wrapped in connective tissue

72. the part of nervous system that consists of the sympathetic and parasympathetic divisions

73. a sensory receptor that detects physical deformations in the body's environment

74. the jellylike material that fills the posterior cavity of the vertebrate eye

75. a sensory receptor that detects heat or cold

77. the small gaps in the myelin sheath of an axon

85. a relay point at the tip of a transmitting neuron's axon

90. the region of the vertebrate ear that conveys vibrations from the eardrum to the oval window

94. a motor neuron and all the muscle fibers it controls

98. the type of adaptation that makes sensory neurons less sensitive when they are stimulated repeatedly

99. the division of the motor nervous system composed of motor neurons that carry signals to skeletal muscles

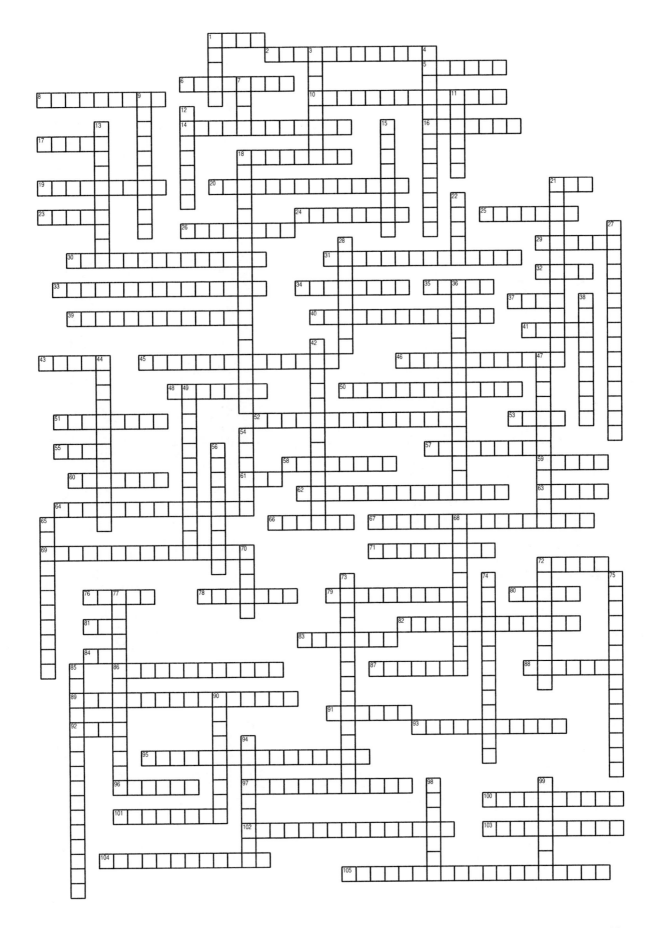

CHAPTER 28

The Life of a Flowering Plant

Studying Advice

a. Have you ever considered how much plants are a part of our daily lives? Look around you right now. How many plant materials do you see? This book page is made of paper, derived from wood. Perhaps you have a pencil nearby with a wooden shaft? You might have a bag of chips or other plant product around for a snack. The curtains or material covering the furniture nearby are likely covered with plant-derived materials (such as cotton), including the clothing you are currently wearing. Plants and their products are all around us. This chapter is about their parts, what they do, and what we do with them. What would we do without them?

b. This chapter should be read before Chapter 29. The figures in this chapter and the organizing tables below are especially helpful in understanding and learning this information.

Student Media

Activities

Roots, Stems, and Leaves
Primary and Secondary Growth
The Life Cycle of a Flowering Plant
Seed and Fruit Formation and Germination

Case Studies in the Process of Science

How Do the Tissues in Monocot and Dicot Stems Compare?
What Tells Desert Seeds When to Germinate?

MP3 Tutors

From Flower to Fruit

Videos

Discovery Channel Video Clip: Plant Pollination

Root Growth in a Radish Seed (time-lapse)

Organizing Tables

Compare the structure of monocots and dicots in the table below.

TABLE 28.1		
	Monocots	Dicots
The number of seed leaves		
Pattern of veins in the leaves		
The number of flower petals and other parts		
Root systems		
Examples		

Compare the structures and functions of the following types of plant cells.

TABLE 28.2		
	Structure	Function
Parenchyma cells		
Collenchyma cells		
Sclerenchyma cells		
Water-conducting cells		
Food-conducting cells		

Compare annuals, biennials, and perennials in the table below.

TABLE 28.3		
	Definition	Examples
Annuals		
Biennials		
Perennials		

Content Quiz

Directions: Identify the *one* best answer for the multiple-choice questions. For true/false questions, determine if the statement is true or false. If false, change the underlined word(s) to make the statement true. Finally, add the correct word(s) to the fill-in-the-blank questions to make the statements true.

Biology and Society: Plant Cloning—Feast and Famine

1. Which one of the following statements is *false*?
   A. Potatoes were introduced to Europe in the 16th century.
   B. Potatoes have been grown in South America for thousands of years.
   C. A potato famine caused Ireland's population to drop by about 25% in just a few years.
   D. A mature potato can be split and planted and each part can develop into a new potato plant that is genetically identical to the parent plant.
   E. A genetically homogenous population is more likely to survive when an environment changes.

2. Which one, if any, of the following does *not* come from plants?
   A. lumber
   B. fabric
   C. paper
   D. industrial chemicals
   E. medicines
   F. All of the above come from plants.

3. True or False?   Plant <u>roots</u> prevent soil erosion.

4. A late blight is a(n) _____-like organism that causes potato rot.

The Structure and Function of a Flowering Plant
MONOCOTS AND DICOTS

5. Which one of the following statements is *false*?
    A. Most monocots have leaves with parallel veins.
    B. Monocots have flowers with petals and other parts in multiples of three.
    C. Monocots have a large, vertical root.
    D. Monocots have stems with vascular tissues arranged in a complex array of bundles.
    E. Monocots include the orchids, palms, lilies, grains, and other grasses.

6. Which one of the following statements is *false*?
    A. Dicots have leaves with a multibranched network of veins.
    B. Dicots typically have a taproot.
    C. Dicots have flowers with petals and other parts in multiples of four or five.
    D. Dicot stems do not have vascular bundles.
    E. Dicots include shrubs, trees, and many of our food crops.

7. True or False?   Dicot embryos typically have <u>one leaf</u>.

8. The first leaves that appear on a plant embryo are _____.

PLANT ORGANS: ROOTS, STEMS, AND LEAVES

9. Which one of the following is *not* an evolutionary adaptation that made it possible for plants to move onto land? The ability to:
    A. use photosynthesis.
    B. absorb light and take in carbon dioxide from the air.
    C. take up water and minerals from the soil.
    D. survive dry conditions.

10. Which one of the following is *not* a function of a plant's root system?
    A. anchors the plant in the soil
    B. produce sugar
    C. transports minerals and water
    D. absorbs minerals and water
    E. stores food

11. True or False?   In a tree, the trunk and branches are its <u>roots</u>.

12. True or False?   Grasses and most other monocots have long leaves <u>without</u> petioles.

13. True or False?   The spines of a barrel cactus are modified <u>stems</u>.

14. Carrots and turnips are examples of _____, which store food.

15. Stems, leaves, and adaptations for reproduction are all part of the _____ system of a plant.

16. In a plant stem, the points at which leaves are attached are called _____, and the portions of the stem in between are called _____.

17. In the process called _____, the terminal bud of a plant produces hormones that inhibit growth of its axillary buds.

18. Removing the terminal buds of some plants stimulates the growth of _____ buds.

19. Using asexual reproduction, a strawberry plant produces a new plant at the tip of its _____.

20. A sweet potato is a modified _____ but a white potato is a modified _____.

21. Stems and leaves depend on the water and minerals absorbed by _____.

22. Located near root tips, tiny projections called _____ increase the surface area for absorption of water and minerals.

23. The primary sites of photosynthesis in most plants are the _____.

24. A sweet pea plant uses a modified leaf called a(n) _____ to climb up its supports.

PLANT CELLS

Matching: Match each item on the left to its best description on the right.

_____ 25. chloroplasts

_____ 26. central vacuole

_____ 27. cell walls

_____ 28. parenchyma cells

_____ 29. collenchyma cells

_____ 30. food-conducting cells

_____ 31. water-conducting cells

_____ 32. sclerenchyma cells

A. provide support in growing parts of plant

B. cell part made mainly of cellulose

C. form chains with overlapping ends, are dead when mature

D. helps maintain turgor in a cell

E. form chains with overlapping ends, are alive when mature

F. organelles that contain photosynthetic pigments

G. their rigid cell walls support the plant like steel beams

H. most abundant type of cell in most plants

33. True or False?   A distinctive feature of plant cells is their <u>vacuole</u> surrounding the plasma membrane.

34. Plant cell walls are made mainly of the structural carbohydrate _____.

35. Long water-conducting cells with tapered ends are called _____; wider, shorter, and less tapered cells are called _____.

PLANT TISSUES AND TISSUE SYSTEMS

Matching: Match each item on the left to its best description on the right.

_____ 36. phloem

_____ 37. xylem

_____ 38. epidermis

_____ 39. cuticle

_____ 40. endodermis

_____ 41. pith

_____ 42. guard cells

_____ 43. mesophyll

_____ 44. cortex

_____ 45. stomata

A. the first defense against physical damage and infections

B. a waxy coating

C. site of food storage and water and mineral absorption in a root

D. regulate the size of the stomata

E. cells that conduct water and dissolved minerals

F. fills the center of the stem in dicots

G. tiny pores between two specialized epidermal cells

H. the ground tissue system of a leaf

I. food-conducting cells

J. the innermost layer of cortex

46. True or False?   <u>All</u> plant stems have vascular tissue systems arranged in numerous vascular bundles.

47. True or False?   Each vein in a leaf is a vascular bundle composed of xylem and phloem surrounded by <u>parenchyma</u> cells.

48. Made up of xylem and phloem, the _____ tissue system provides support and transports water and nutrients throughout the plant.

49. Filling the spaces between the epidermis and vascular tissue system, the _____ system makes up the bulk of a young plant.

50. The _____ tissue system of plants covers and protects leaves, stems, and roots.

Plant Growth

51. A plant that lives for two years, typically flowering in the second year, is:
    A. an annual.
    B. a biennial.
    C. a biannual.
    D. a perennial.
    E. a centennial.

52. True or False?   Organisms that grow throughout their lives show <u>determinate</u> growth.

53. Plants that live and reproduce for many years are called _____.

54. Plants that live and reproduce in just one year or growing season are called _____.

PRIMARY GROWTH: LENGTHENING

55. Which one of the following statements about meristems is *false*?
    A. Meristems consist of unspecialized cells that divide and generate new cells and tissues.
    B. Meristems are present only during the embryonic stages of a plant's life.
    C. Cell division in the apical meristems of roots and shoots contributes to primary growth.
    D. Growth in all plants is made possible by meristems.

56. True or False?   Cells produced by <u>secondary</u> growth form tissues that develop into the epidermis, cortex, and vascular tissue.

57. At the tip of a root is a(n) _____, a thimble-like cone of cells that protects the apical meristem.

58. Meristems at the tips of roots and in the terminal and axillary buds of shoots are called _____.

59. Cell division in the apical meristems of roots and shoots produces _____ growth.

SECONDARY GROWTH: THICKENING

60. Secondary growth involves cell division in two meristems,
    A. the vascular cambium and the cork cambium.
    B. the vascular cambium and the apical meristems.
    C. the cork cambium and the apical meristems.
    D. wood rays and apical meristems.
    E. the cork cambium and rays.

61. Which one of the following statements about growth rings is *false?*
    A. Annual growth rings result from layers of uneven secondary xylem growth due to seasonal variations.
    B. Early wood cells are usually smaller in diameter and thicker-walled than those produced in summer.
    C. Each tree ring consists of a cylinder of early wood surrounded by a cylinder of summer wood.
    D. The vascular cambium becomes dormant each year during winter.

62. After several decades of growth by a tree, which one of the following is typically dead?

    A. the vascular cambium

    B. the cork cambium

    C. the youngest secondary phloem

    D. mature cork cells

    E. cells in the wood rays

63. True or False? Stems and roots often thicken through primary growth.

64. True or False? After several decades of secondary growth, the bulk of a tree trunk is dead tissue.

65. During secondary growth, the _____ first appears as a cylinder of actively dividing cells between the primary xylem and primary phloem.

66. Yearly production of a new layer of secondary _____ accounts for most of the growth in thickness of a perennial plant.

67. Secondary xylem consists of xylem cells and fibers that have thick walls rich in _____.

68. Cork is produced by a meristem called _____.

69. Everything external to the vascular cambium (the secondary phloem, cork cambium, and cork) is called _____.

The Life Cycle of a Flowering Plant

THE FLOWER

Matching: Match each item on the left to its best description on the right.

_____ 70. petals

_____ 71. stamens

_____ 72. ovule

_____ 73. sepals

_____ 74. carpel

_____ 75. ovary

_____ 76. style

_____ 77. anther

_____ 78. stigma

A. the female organ of the flower

B. the receiving surface for pollen grains

C. the male organs

D. it leads to the ovary at the base of the stigma

E. often bright and colorful they advertise a flower to pollinators

F. contains the developing egg and cells that support it

G. a terminal sac on the stamen

H. enclose and protect the flower bud

I. houses reproductive structures called the ovules

79. True or False? A flower's main parts, the sepals, petals, stamens, and carpels, are modified stems.

80. True or False?   Many plants can reproduce <u>sexually and asexually</u>.

81. True or False? Using <u>sexual</u> reproduction, a single plant can produce many identical plants quickly and efficiently.

82. In angiosperms, the organ specific to sexual reproduction is the _____.

83. Fertilization occurs in the _____.

POLLINATION AND FERTILIZATION

Matching: Match each item on the left to its best description on the right.

_____ 84. sporophyte

_____ 85. gametophyte

_____ 86. pollen grain

_____ 87. embryo sac

_____ 88. pollination

_____ 89. fertilization

A. a plant's haploid generation

B. the female gametophyte of angiosperms

C. structure that contains immature male gametophytes

D. the diploid plant body

E. the delivery of pollen to the stigma of a carpel

F. the union of male and female gametes

90. True or False?   The pollen grain germinates on the <u>stamen</u>.

91. The formation of a zygote and a cell with a triploid nucleus is called _____.

SEED FORMATION

92. After a mature seed is produced,
    A. the endosperm disintegrates.
    B. the seed coat splits open.
    C. the zygote divides by meiosis.
    D. the embryo stops developing.
    E. an embryonic root emerges.

93. True or False?   The <u>cotyledon</u> encloses and protects the endosperm.

94. The _____ is a multicellular mass that nourishes the embryo until it becomes a self-supporting seedling.

FRUIT FORMATION, SEED GERMINATION

95. Which one of the following statements about fruits is *false?*
    A. Fruits are highly variable.
    B. A corn kernel is a fruit.
    C. A pea pod is a fruit.
    D. Fruits develop after the seeds develop.
    E. A fruit houses and protects seeds and helps disperse them from the parent plant.

96. Germination is typically triggered by:
    A. mechanical damage to the seed coat.
    B. fertilization.
    C. destruction of the endosperm.
    D. ingestion by an animal.
    E. the absorption of water by the seed.

97. True or False?   The first thing to emerge from a germinating seed is the embryonic <u>shoot</u>.

98. True or False?   In the wild, <u>only a few</u> seedlings endure long enough to reproduce.

99. Fleshy, edible fruits entice animals that help spread _____.

Evolution Connection: The Interdependence of Angiosperms and Animals

100. Which one of the following do angiosperms *not* rely upon for pollination and seed dispersal?
    A. insects
    B. fish
    C. mammals
    D. birds

101. True or False?   Flowers that are pollinated by <u>birds</u> often have markings that reflect ultraviolet light.

102. A high-energy fluid that plants use only for attracting pollinators is called _____.

Word Roots

a = without (asexual reproduction: reproduction without sex)

apic = the tip; **meristo** = divided (apical meristems: embryonic plant tissue on the tips of roots and in the buds of shoots that supplies cells for the plant to grow)

bienn = every two years (biennial: a plant that requires two years to complete its life cycle)

chloro = green (chloroplasts: the green, photosynthetic organelle common in plants)

coll = glue; **enchyma** = an infusion (collenchyma cells: a flexible plant cell type that occurs in strands or cylinders that support young parts of the plant without restraining growth)

di = two (dicot: a flowering plant whose embryos have two seed leaves)

endo = inner; **derm** = skin (endodermis: the innermost layer of the cortex in plant roots)

epi = over (epidermis: the dermal tissue system in plants; the outer covering of animals)

gamet = a wife or husband (gametophyte: the multicellular haploid form in organisms undergoing alternation of generations, which mitotically produces haploid gametes that unite and grow into the sporophyte generation)

inter = between (internode: the segment of a plant stem between the points where leaves are attached)

meso = middle; **phyll** = a leaf (mesophyll: the ground tissue of a leaf, sandwiched between the upper and lower epidermis and specialized for photosynthesis)

mono = one (monocot: a flowering plant whose embryos have a single seed leaf)

perenni = through the year (perennial: a plant that lives for many years)

phloe = the bark of a tree (phloem: the portion of the vascular system in plants consisting of living cells arranged into elongated tubes that transport sugar and other organic nutrients throughout the plant)

sclero = hard (sclerenchyma cells: a supportive cell with a very rigid secondary wall)

sporo = a seed; **phyto** = a plant (sporophyte: the multicellular diploid form in organisms undergoing alternation of generations that results from a union of gametes and that meiotically produces haploid spores that grow into the gametophyte generation)

stam = standing upright (stamen: the pollen-producing male reproductive organ of a flower, consisting of an anther and filament)

stoma = mouth (stomata: a hole in plant leaves through which water, oxygen, and carbon dioxide pass)

xyl = wood (xylem: the tube-shaped, nonliving portion of the vascular system in plants that carries water and minerals from the roots to the rest of the plant)

vascula = little tube (vascular tissue system: a transport system of tubes for water and nutrients in plants)

Key Terms

angiosperms
annuals
anther
apical dominance
apical meristem
asexual
  reproduction
axillary buds
bark
biennials
blade
carpel
cell wall
central vacuole
chloroplasts
collenchyma cells
cork
cork cambium
cortex
cotyledons
cuticle
dermal tissue
  system

determinate
  growth
dicot
double
  fertilization
embryo sac
endodermis
endosperm
epidermis
fertilization
filament
food-conducting
  cells
fruit
gametophyte
germinates
ground tissue
  system
guard cells
indeterminate
  growth
internodes
leaves

lignin
meristem
mesophyll
monocot
nodes
ovary
ovule
parenchyma cells
perennials
petals
petiole
phloem
pith
pollen grain
pollination
primary cell wall
primary growth
root cap
root hairs
root system
sclerenchyma cells
secondary cell wall

secondary growth
seed coat
sepals
shoot system
sporophyte
stamen
stems
stigma
stomata
style
terminal bud
tissue systems
tracheids
tubers
vascular cambium
vascular tissue
  system
vessel elements
water-conducting
  cells
wood
xylem

Crossword Puzzle

Use the Key Terms list from this chapter to fill in the crossword puzzle.

ACROSS

2. unspecialized cells that divide and generate new cells and tissues

4. a mature ovary of a flower that protects dormant seeds and aids in their dispersal

7. the multicellular haploid form in organisms undergoing alternation of generations

11. the ground tissue of a leaf

12. what a seed does when it first begins to grow

14. a point along the stem of a plant where a leaf is attached

15. the type of tissue system consisting mostly of parenchyma cells that makes up the bulk of a young plant

16. a subdivision of flowering plants whose members possess two embryonic seed leaves

19. the union of haploid gametes to produce a diploid zygote

20. a type of system formed by xylem and phloem throughout a plant, serving as a transport system for water and nutrients

21. a chemical component of wood

22. the female gametophyte contained in the ovule of a flowering plant

23. plant tissues are organized into these

24. the type of tissue system that covers and protects leaves, stems, and roots

27. a continuous cylinder of meristematic cells surrounding the xylem and pith that produces secondary xylem and phloem

29. sites of photosynthesis in plant cells

32. a waxy coating on the surface of stems and leaves that helps retain water

33. the type of fertilization in angiosperms in which two sperm cells unite with two cells in the embryo sac to form the zygote and endosperm

35. a tough protective layer that encloses the endosperm

36. the ground tissue system of a root, which stores food and absorbs minerals

37. a plant that requires two years to complete its life cycle

39. all of a plant's roots that anchor it in the soil, absorb and transport minerals and water, and store food

42. the outermost protective layer of a plant's bark

45. seed leaves

46. the type of meristem found at the tips of roots and in the terminal and axillary buds of shoots

47. the pollen-producing male reproductive organ of a flower

49. meristematic tissue that produces cork cells during secondary growth of a plant

51. a plant that completes its entire life cycle in a single year or growing season

52. the type of cell wall deposited between the plasma membrane and primary cell wall

53. the type of growth that ends after an organism reaches a certain size

55. the tube-shaped, nonliving portion of the vascular system in plants that carries water and minerals

57. the type of specialized epidermal plant cells that regulate the size of stomata

58. the type of reproduction that creates offspring derived from a single parent

60. the aerial portion of a plant body, consisting of stems, leaves, and flowers

61. the starch storing portion of a potato plant

62. wider and shorter, open-ended, water-conducting cells

63. a whorl of modified leaves that enclose and protect the flower bud before it opens

65. an embryonic shoot present in the angle formed by a leaf and stem

66. all of the plant tissues external to the vascular cambium

68. a microscopic pore surrounded by guard cells in the epidermis of leaves and stems

70. the female reproductive organ of a flower

71. a part of a plant's shoot system that supports the leaves and reproductive structures

72. in a seed plant, the male gametophytes that develop within the anthers of stamens

73. the type of cell wall laid down first

75. the placement of pollen onto the stigma of a carpel by wind or animal carriers

76. a tiny projection growing just behind a root tip, it increases the surface area of a root

DOWN

1. the main site of photosynthesis in most plants

3. the two types of cells that form a chain of water-carrying tubes

4. a type of specialized cell that forms phloem tissue

5. a part of the ground tissue system that fills the center of a stem and is often important in food storage

6. the type of growth characteristic of plants, in which the organism grows as long as it lives

8. the stalk of a stamen

9. the innermost layer of the cortex in plant roots

10. the skin of a plant that covers and protects leaves, young stems, and young roots

13. a plant that lives for many years

17. a large fluid-containing organelle in plant cells

18. an embryonic tissue at the tip of a shoot

25. a modified leaf of a flowering plant that is often the most colorful part of a flower

26. a supportive structure that surrounds plant cells

28. the stalk of a leaf, which joins the leaf to the stem

29. the type of flexible plant cell type that occurs in strands or cylinders that support young parts of the plant without restraining growth

30. in flowers, the portion of a carpel in which the egg-containing ovules develop

31. the segment of a plant stem between the points where leaves are attached

34. type of growth characterized by an increase in girth of the stems and roots

38. the type of dominance in which the terminal bud produces hormones that inhibit growth of axillary buds

40. when mature, these types of cells are dead, and their rigid cell walls support the plant much like steel beams in a building

41. a subdivision of flowering plants whose members possess one embryonic seed leaf

43. vascular tissue that contains food-conducting cells

44. long and narrow, porous, water-conducting cells

48. the general name for flowering plants

50. a cone of cells at the tip of a plant root that protects the apical meristem

54. a nutrient-rich tissue, which provides nourishment to the developing embryo in angiosperm seeds

56. a structure that develops in the plant ovary and contains the female gametophyte

59. the terminal pollen sac of a stamen

60. the sticky part of a flower's carpel

63. the diploid plant body

64. the flattened portion of a leaf

67. the stalk of a flower's carpel, with the ovary at the base and the stigma at the top

69. the most abundant type of cell in most plants

74. the type of growth initiated by the apical meristems of a plant root or shoot

The Working Plant

Studying Advice

a. Have you ever wondered how seedless grapes are produced? Have you ever struggled to keep a plant alive, wondering what you were doing wrong? Have you ever wondered how plants grow towards sunlight? This chapter reveals why and how plants do what they do and makes caring for plants a little easier.

b. Chapter 29 builds upon the content in Chapter 28. If you have not already read Chapter 28, be prepared to read at least sections of it as background for Chapter 29.

Student Media

Activities

Absorption of Nutrients from Soil

Transpiration

Transport in Phloem

Leaf Abscission (Drop)

Flowering Lab

Case Studies in the Process of Science

How Does Acid Precipitation Affect Mineral Deficiency?

What Determines if Water Moves Into or Out of a Plant Cell?

How Is the Rate of Transpiration Calculated?

What Plant Hormones Affect Organ Formation?

Graph It

Global Soil Degradation

MP3 Tutors

Transpiration

Videos

Phototropism
Gravitropism
Mimosa Leaf
Sun Dew Trapping Prey

You Decide

Is Ephedra Safe and Effective?

Organizing Tables

Distinguish between micronutrients and macronutrients in the table below.

TABLE 29.1		
	Definition	Examples
Micronutrients		
Macronutrients		

Identify the main functions of the five plant hormones in Table 29.2 below.

TABLE 29.2		
	Plant Hormones	Functions
Auxin		
Ethylene		
Cytokinins		
Gibberellins		
Abscisic Acid		

Distinguish between the three types of tropisms in the table below.

TABLE 29.3

	Definition	Examples
Phototropism		
Thigmotropism		
Gravitropism		

Content Quiz

Directions: Identify the *one* best answer for the multiple-choice questions. For true/false questions, determine if the statement is true or false. If false, change the underlined word(s) to make the statement true. Finally, add the correct word(s) to the fill-in-the-blank questions to make the statements true.

Biology and Society: What's in—or Not in—That Organic Tomato?

1. Which one of the following is *not* characteristic of organic farming?
   A. maintaining and replenishing soil quality
   B. the use of natural pest parasites
   C. crop rotation
   D. extensive use of fertilizers
   E. protection of biological diversity

2. True or False?   Scientists <u>agree</u> that organic foods are better than conventional crops.

3. A guiding principle of _____ farming is that the nutritional health of plants affects the nutritional health of those who eat the plants.

How Plants Acquire and Transport Nutrients
PLANT NUTRITION

4. Which one of the following is a micronutrient?
   A. carbon
   B. oxygen
   C. hydrogen
   D. nitrogen
   E. iron

5. Which one of the following is a macronutrient?
   A. manganese
   B. copper
   C. sulfur
   D. chlorine
   E. zinc

6. A plant needs only minute quantities of micronutrients because it:
   A. can easily get them from the soil.
   B. uses micronutrients over and over.
   C. can produce them from macronutrients.
   D. really needs them only a few days of the year.
   E. needs them only to reproduce.

7. True or False?   Plants require relatively large amounts of <u>micronutrients</u>.

8. True or False?   It is <u>relatively easy</u> to diagnose nutrient deficiencies in plants.

9. True or False?   A deficiency of any micronutrient <u>can kill</u> a plant.

10. An element is considered a(n) _____ plant nutrient if the plant must obtain it to complete its life cycle.

FROM THE SOIL INTO THE ROOTS

11. Which one of the following statements about how plants obtain nutrients is *false*?
    A. A plant's source of carbon is atmospheric CO_2.
    B. A plant's source of oxygen is atmospheric CO_2.
    C. A plant's source of hydrogen is water.
    D. To reach the xylem, soil solutions must pass through the plasma membranes of root cells.
    E. A plant absorbs all of its essential nutrients from the soil.

12. True or False?   <u>All</u> substances that enter a plant root are dissolved in water.

13. Roots efficiently extract nutrients from soil due to their _____, extensions of epidermal cells that dramatically increase the surface available for absorption.

14. Most plants gain absorptive surface through mutually beneficial symbiotic associations with fungi in an association called _____.

15. The selectively permeable plasma membrane of _____ cells regulates what solutes reach the xylem.

THE ROLE OF BACTERIA IN NITROGEN NUTRITION

Matching: Match each item on the left to its best description on the right.

_____ 16. nitrogen-fixing bacteria

_____ 17. ammonifying bacteria

_____ 18. nitrifying bacteria

_____ 19. legumes

_____ 20. root nodules

_____ 21. nitrogen fixation

A. add ammonium to soil by decomposing organic matter

B. bacteria that convert atmospheric N_2 to ammonium

C. plants that produce their seeds in pods

D. the process of converting atmospheric N_2 to ammonium

E. convert soil ammonium to nitrate

F. plant root cells containing vesicles filled with bacteria

22. True or False?   Plants <u>can</u> use the form of nitrogen found in air.

23. For plants to absorb _____ from the soil, it must first be converted to ammonium ions (NH_4^+) or nitrate ions (NO_3^-).

THE TRANSPORT OF WATER

24. Transpiration exerts a pull on a tense string of water molecules that is held together by _____ and helped upward by _____.
    A. cohesion, adhesion
    B. adhesion, cohesion
    C. cohesion, gravity
    D. adhesion, gravity
    E. gravity, cohesion

25. Which one, if any, of the following environmental conditions would *not* increase the amount of transpiration in a plant?
    A. intense sunlight
    B. warm weather
    C. high humidity
    D. windy conditions
    E. All of the above would increase the amount of transpiration in a plant.

26. Which one of the following statements is *false*?
    A. Stomata are usually open during the day.
    B. Stomata are usually open at night to allow oxygen to enter from the atmosphere.
    C. During the day, CO_2 can enter the leaf from the atmosphere.
    D. Stomata may also close during the day if a plant is losing water too fast.
    E. Two guard cells flanking each stoma control its opening by changing shape.

27. True or False?   The sticking together of molecules of the same kind defines adhesion.

28. True or False?   The transport of xylem sap requires <u>no energy</u> expenditure by the plant.

29. Mature water-conducting cells of _____ conduct xylem sap from a plant's roots to the tips of its leaves.

THE TRANSPORT OF SUGARS

30. Which one of the following statements about the pressure-flow mechanism is *false*?
    A. Each food-conducting tube in phloem tissue has a source end and a sink end.
    B. At source, sugar is loaded from a photosynthetic cell into a phloem tube.
    C. At the sugar sink, water and sugar enter the phloem tube.
    D. At the sugar sink, the exit of water lowers the water pressure in the tube.
    E. Phloem sap always flows from a sugar source to a sugar sink.

31. True or False?   A sugar <u>sink</u> is a location in a plant where sugar is being produced.

32. True or False?   <u>Phloem sap</u> moves freely from the cytoplasm of one cell to the next.

33. The main function of _____ is to transport sugars produced by a plant using photosynthesis.

Plant Hormones

AUXIN, ETHYLENE, CYTOKININS, GIBBERELLINS, ABSCISIC ACID

Matching: Match each hormone on the left to its best description on the right.

_____ 34. auxin   A.   promotes seed germination

_____ 35. ethylene   B.   promotes cell elongation in stems

_____ 36. cytokinin   C.   promotes fruit ripening and dropping of leaves

_____ 37. gibberellin   D.   inhibits seed germination

_____ 38. abscisic acid E.   promotes cytokinesis

39. Which one of the following statements about plant hormones is *false*?
    A. Plants produce hormones in very large amounts.
    B. Each type of hormone can produce a variety of effects.
    C. Small amounts of hormones can have profound effects on target cells.
    D. Hormones play critical roles in the development of plants.
    E. Hormones play critical roles in the growth of plants.

40. Which one of the following statements about auxin is *false?*
    A. Auxin promotes cell elongation in stems only within a certain concentration range.
    B. Auxin is produced by developing seeds and promotes the growth of fruit.
    C. The same concentration of auxin that promotes cell elongation in stems inhibits cell elongation in roots.
    D. Auxin is the hormone most associated with the loss of leaves in autumn.
    E. Auxin promotes growth in stem diameter.

41. Which one of the following hormones enters the shoot system from the roots and counters the inhibitory effects of auxin coming down from the terminal buds?
    A. gibberellins
    B. auxin
    C. ethylene
    D. cytokinins
    E. None of the above.

42. Which of the following pairs stimulate cell elongation and cell division in stems and influence fruit development?
    A. gibberellins and auxin
    B. cytokinins and ethylene
    C. abscisic acid and ethylene
    D. auxin and abscisic acid
    E. cytokinins and abscisic acid

43. True or False?   An uneven distribution of auxin causes the dark side of a shoot to grow <u>slower</u> than the light side.

44. True or False?   The same hormone <u>may have different</u> effects at different concentrations in the same target cell.

45. True or False?   Animals <u>and plants</u> use hormones to regulate their internal activities.

46. True or False?   Some growers retard ripening by storing apples in containers flushed with <u>ethylene</u>.

47. Regulatory chemicals that travel from their production sites to affect other parts of the body are called _____.

48. The hormone _____ promotes cell elongation in stems.

49. The primary internal signal that enables plants to withstand drought is _____.

50. Spraying unfertilized plants with synthetic _____ can produce seedless tomatoes, cucumbers, and eggplants.

51. The growth of a shoot toward light is called _____.

Response to Stimuli
TROPISMS, PHOTOPERIODS

52. Long-night plants respond to:
    A. warmer daytime temperatures.
    B. stronger intensities of light.
    C. the relative length of time without sunlight.
    D. phases of the moon.
    E. the total amount of sunlight through the day.

53. True or False?   Short-night plants usually flower in late <u>fall or early winter</u>.

54. Growth of a plant in response to touch is called _____.

55. Growth of a plant in response to gravity is called _____.

Evolution Connection: The Interdependence of Organisms

56. Most flowering plants depend on:
    A. insects or other animals for pollination and seed dispersal.
    B. nitrogen-fixing bacteria.
    C. the fungi of mycorrhizae.
    D. the sun for photosynthesis.
    E. All of the above.

57. True or False?   The fossil record shows that <u>mycorrhizae</u> have existed since plants first evolved.

58. A needle-like mouthpart is used by _____ to feed on plant sap.

59. The praying mantis pictured on this book's cover feeds on _____.

Word Roots

aux = grow, enlarge (auxins: a class of plant hormones that promote cell elongation, stimulate secondary growth, and promote the development of leaf traces and fruit)

co = together; **hesita** = stick fast (cohesion: the sticking together of molecules of the same kind)

cyto = cell; **kine** = moving (cytokinins: a class of related plant hormones that retard aging and act in concert with auxins to stimulate cell division, influence the pathway of differentiation, and control apical dominance)

gibb = humped (gibberellins: a class of related plant hormones that stimulate growth in the stem and leaves, trigger the germination of seeds and breaking of bud dormancy, and stimulate fruit development with auxin)

gravi = heavy; **trop** = turn, change (gravitropism: directional growth of plants in response to gravity)

macro = large (macronutrients: elements required by plants and animals in relatively large amounts)

micro = small (micronutrients: elements required by plants and animals in very small amounts)

myco = a fungus (mycorrrhizae: mutualistic associations of plant roots and fungi)

photo = light (phototropism: growth of a plant shoot toward or away from light)

thigmo = touch (thigmotropism: growth of a plant in response to touch)

Key Terms

abscisic acid (ABA)	gravitropism	photoperiod	transpiration
adhesion	hormones	phototropism	transpiration-
auxin	macronutrients	pressure-flow	cohesion-tension
cohesion	micronutrients	mechanism	mechanism
cytokinins	minerals	root nodules	tropisms
essential element	mycorrhiza	sugar sink	xylem sap
ethylene	nitrogen fixation	sugar source	
gibberellins	phloem sap	thigmotropism	

Crossword Puzzle

Use the Key Terms list from this chapter to fill in the crossword puzzle.

ACROSS

1. the type of mechanism by which phloem sap is transported through a plant from a sugar source to a sugar sink
2. growth of a plant shoot toward or away from light
3. the directional growth of a plant in relation to touch
7. the assimilation of atmospheric nitrogen by certain prokaryotes into nitrogenous compounds that can be directly used by plants
8. the abbreviation for the plant hormone that inhibits the germination of seeds and promotes dormancy
10. a mutualistic association of a plant root and fungus
11. the evaporative loss of water from a plant
12. a plant organ that is a net consumer or storer of sugar
13. swellings consisting of plant cells that contain nitrogen-fixing bacteria
16. organic ions
17. the type of nutrients that plants require in relatively large amounts
18. the sticking together of molecules of the same kind
19. a type of element that a plant must obtain to complete its life cycle
22. regulatory chemicals that travel away from their site of origin to affect internal activities
23. growth of a plant in response to gravity
25. any chemical substance that promotes seedling elongation
26. a family of more than 100 growth-regulating plant hormones

DOWN

1. the relative lengths of day and night that allow plants to detect the time of year
4. a growth response that makes a plant grow toward or away from a stimulus
5. the type of nutrients that plants require in extremely small amounts
6. the type of mechanism used to describe the movement of xylem sap
9. the sticking together of molecules of different kinds
14. a plant organ in which sugar is being produced
15. growth regulators that promote cytokinesis
20. a hormone that triggers a variety of aging responses in plants, including fruit ripening and dropping of leaves
21. a sugary solution that moves throughout a plant in various directions
24. a solution of mostly inorganic nutrients that flow through vertical tubes from the roots to the tips of the leaves